From Fibres to

Elizabeth Ga

Tutor in Weaving, Stanhope Institute of Education

Mills & Boon Limited, London

First published 1968 by Allman & Son Ltd, 17–19 Foley Street,
London W1A 1DR

Second edition 1971
Third edition 1978 published by Mills & Boon Limited
Reprinted 1980

© Elizabeth Gale 1968
New material in this edition, © Elizabeth Gale 1978

ISBN 0 263 06376 3

Printed in Great Britain by
M & A Thomson Litho Ltd., East Kilbride and
bound by Hunter & Foulis Ltd., Edinburgh

Contents

LIST OF PLATES		vii
ACKNOWLEDGEMENTS		viii
PREFACE		ix
1	NATURAL YARNS: GROWTH AND MANUFACTURE OF VEGETABLE FIBRES	1

    INTRODUCTION
    Fibres; Union Cloths; Staple and Filament; Natural Yarns; Cellulosic; Proteinic; Cotton; Linen; Jute; Hemp; Ramie; Coir; Sisal; Kapok; Paper Yarns

2	NATURAL YARNS: SOURCE AND MANUFACTURE OF ANIMAL FIBRES	11

    Wool; Classification of Breeds; Re-Manufactured wool; Worsted; Mohair; Camel hair; Angora; Cashmere; Vicuna; Guanaco; Alpaca; Llama; Silk

3	MAN-MADE YARNS: ORIGIN AND PRODUCTION	21

    Synthetic: Cellulosic; Proteinic; Alginate; Mineral. Polyamide Fibres: Nylon; Bri-Nylon; Celon; Perlon; Enkalon; Fluflon; Miralon. Polyester Fibres: Terylene; Crimplene; Dacron; Fluflene; Miralene. Acrylic Fibres: Acrilan; Courtelle; Orlon; Dralon. Modacrylic Fibres: Teklan; Vinyon. Polythene Fibres: Courlene. Polyuretha
    Spanzelle. Polypropylene Fibres: Ulstron. Cellulosic Fibres: Viscose Rayon; Vincel; Evlan; Durafil; Sarille; Fibro; Rayolanda. Cellulose Ester Fibres: Acetate Rayon; Dicel; Tricel; Tricelon; Celafibre; Cuprammonium Rayon. Proteinic Fibres: Fibrolane; Ardil; Vicara; Alginate. Mineral Fibres: Glass; Asbestos.

4	SPINNING METHODS	37

    Flyer Spinning; Ring Spinning; Cap Spinning; Mule Spinning. Folded or plied Yarn; Twist of Yarn; Fancy or Novelty Yarns; Spiral Yarn; Gimp Yarn; Snarl Yarn;

Knop Yarn; Cloud Yarn; Tuffle Yarn; Nep Yarn; Curl or Loop Yarn; Boucle; Grandelle Yarn; Marl Yarn; Nub Yarn; Flake Yarn; Chenille Yarn; Lurex. The Size of Yarns; Simple Tests to Identify Yarns; Burning Tests; Caustic Soda Tests

5 THE DYEING OF YARN

Hanks; Cheeses of Yarn; Warp Beam; Cake and Cop

6 WOVEN FABRICS

The Sett of Warp and Weft; Warping; Weaving of Fabrics; Drafts and Peg Plans; Plain Weave; Hopsack. Analysis of Cloth for Draft and Peg Plan

7 FABRICS PRODUCED WITH PLAIN WEAVE

Appliqued Fabric; Azlin; Bag Cloths; Baize; Batiste; Billiard Cloth; Broadcloth; Buckram; Calico; Canton; Cambric; Camlet; Casement Cloth; Cheese Cloth; Chiffon; Crash; Crepe de Chine; Duck; Georgette; Gingham; Glass Cloth; Gloria; Grosgrain; Hessian; Holland; Jaconet; Lawn; Limbric; Mull; Muslin; Nainsook; Ninon; Organdie; Ottoman; Petersham; Pina Cloth; Pongee; Poplin; Pyjama Cloth; Repp; Scrim; Seersucker; Sheeting; Shantung; Taffeta; Tussore; Voile

8 FABRICS IN TWILL, SATIN AND SATEEN WEAVES

Barathea; Blankets; Botany Twill Cloths; Box Cloth; Coutil; Denim; Drills; Dungaree; Estamene; Felt (Woven); Flannel; Flannelette; Frieze; Foulard; Gabardine; Galatea; Habutai; Imperial Cloth; Kersey; Linsey, Maud; Melton Cloth; Serge; Tartan; Ticking; Tweed; Twills; Herringbone and Chevron; Satin; Sateen. Fabrics in Satin and Sateen Weaves: Amazon; Atlas; Beaver Cloth; Dimity; Doeskin; Duchesse Satin; Habit Cloth; Japanese Satin; Luvisca; Madras Shirting; Satinet; Silesia; Venetian

9   FABRICS WOVEN ON THE DOBBY LOOM   90
        Bedford Cord; Distorted Weft; Double Cloth; Folkweave;
        Honeycomb; Huckaback; Spot Designs; Pique

10 FABRICS WOVEN ON THE JACQUARD LOOM   101
        The Cards; The Loom; The Fabrics; Damask; Brocade;
        Lampas; Brocatelle; Tapestry; Matelasse

11 PILE FABRICS   119
        Velvet; Genoa Velvet; Terry Velvet; Utrecht Velvet;
        Figured Velvet; Velour; Velveteen; Figured Velveteen;
        Corduroy; Needlecord; Designs on Plain Warp Pile
        Fabrics; Plush; Moquette; Towelling

12 NET CURTAINING AND VISION NETS   130
        Leno and Gauze; Marquisette; Madras Muslin; Gossamer;
        Vision Nets; Swivel and Lappet Weaving

13 PLASTICS AND NON-WOVEN FABRICS   136
        Plastic (P.V.C.) Sheeting; Laminated Plastics; Plastic
        Coated Yarns; Plastic Filament; Bonded Fabrics; Spun
        Bonds; Felt; Paper Fabrics

14 FINISHING OF FABRICS   141
        Desizing; Scouring; Bleaching; Stoving; Mercerisation;
        Raising; Nap Finish; Dress Face Finish; Cropping or
        Shearing; Brushing and Steaming; Singeing; Beetling;
        Crabbing; Milling; Stentering; Waterproofing; Velanizing;
        Crease Resistance; Mothproofing; Rot Proofing; Flame
        Resistance; Sanforizing; Calendering; Schreiner Calender;
        Moiré Fabrics; Embossing; Velvet and Velveteen Finishing;
        Pilling

15 THE PRINTING OF TEXTILES   155
        Block Printing; Screen Printing; Machine Screen Printing;
        Rotary Screen Printing; Duplex Printing; Transfer Printing;
        Dye Mixing for Printing

16   PRINTED AND DYED STYLES
        Direct Printing; Blotch; Mordant; Discharge Style; Resist or Reserve Styles; Batik; Tie and Dye; Printed Warp; Chintz; Cretonne

17   PREPARATION AND PRINTING OF FABRICS
        Cotton; Linen; Wool; Silk; Viscose Rayon; Acetate Rayon; Nylon; Terylene

18   PREPARATION AND DYEING OF FABRICS
        Cotton; Linen; Jute; Hemp; Sisal; Ramie; Wool; Silk; Viscose Rayon; Acetate Rayon; Nylon; Terylene; Crimplene; Orlon; Acrilan; Ardil; Piece Dyed Cloth; Cross-Dyed Cloth

19   DYES
        Classification of Dyes: Direct Dyes; Acid Dyes; Vat Dyes; Basic Dyes; Mordant Dyes; Sulphur Dyes; Pre-Metallised Dyes; Azoic Dyes; Mineral and Pigment Dyes; Oxidation Dyes; Dispersed Dyes

20   THE CARE OF FABRICS AND GARMENTS
        Washing; Starching; Drying; Ironing. Natural Fibres: Cotton; Linen; Wool; Cashmere; Silk. Man-made Fibres: Viscose Rayon; Durafil; Sarille; Vincel; Acetate Rayon; Dicel; Tricel; Nylon; Bri-Nylon; Celon; Enkalon; Terylene; Crimplene; Fibrolane; Courtelle; Acrilan; Teklan; Glass Fibre. Laundry; Dry Cleaning. Removal of Stains: Mildew; Grass Stains; Scorch; Blood Stains; Fruit Stains; Coffee or Tea; Ink; Paint or Varnish; Grease or Oil; Wine; Rust (Iron Mould). Dry Cleaning of Stains

21   KNITTED FABRICS
        Knitting stitches; Types of Needles; Knitting Machines; Types of Fabric

APPENDIX   SHUTTLELESS LOOMS

INDEX

vi

List of Plates

1. Fancy Yarns: (a) Loop Yarn, (b) Snarl Yarn, (c) Gimp Yarn, (d) Gimp Yarn, (e) Flake Yarn, (f) Tuffle Yarn. — 47

2. Fabric woven on the dobby loom from draft and peg plan at Diag. 29. — 94

3. Section of a design on Jacquard point paper in satin and sateen weaves. — 102

4. Furnishing fabric woven on the Jacquard loom from the design in photograph 3. Warp—2/12's black cotton. Weft—gimp yarn. — 103

5. Section of a design on Jacquard point paper for the fabric in photograph 6. — 108

6. Dress Fabric. Warp—1 end gold Lurex, 3 ends yellow nylon. Weft—orange, yellow and fuchsia rayon. Weave—sateen. — 109

7. Damask. — 112
   Width of material 48"/50". Repeat $10\frac{1}{2}$" length, $6\frac{1}{4}$" width. Woven in 69% cotton, 31% silk.

8. Brocatelle. — 114
   Width of material 51"/52". Repeat $21\frac{1}{2}$" length, $25\frac{1}{2}$" width. Woven in 42% linen, 31% silk, 25% rayon, 2% nylon.

9. Tapestry. — 116
   Width of material 50"/51". Repeat 16" length, $12\frac{1}{2}$" width. Woven in 100% cotton.

10. Matelasse. — 117
    Width of material 49"/50". Repeat 6" length, 6" width. Warp—218 ends per inch 150d. viscose 15 turns per inch. Weft—76 picks per inch 30's lea irregular spun viscose slub.

11 Figured Velvet.
Width of material 52"/54". Repeat 16" length, 13½" width. Woven in 60% rayon, 40% cotton, with cut and uncut pile.

12 Batik.
Screen print in twelve colours on 100% cotton.
Width of material 48"/50". Repeat 35½" length and full width of material.

13 Full double jersey 18 gauge knitting machine.

14 Cylinder needle bed showing needle butts.

15 Patterning unit for double jersey machine.

Acknowledgements

I am very grateful to my husband for the excellent diagrams and photographs.

I wish to thank Marianne Straub for her helpful suggestions regarding the woven textiles and Miriam A. Parker-Bright for her help and advice on the knitted fabrics.

I am greatly indebted to Sekers Fabrics Limited for the help and kindness they extended to me at Whitehaven and for permission to use their fabrics in plates 5, 6 and 10.

My thanks to Arthur Sanderson & Sons Ltd. who kindly gave permission to use their fabrics in plates 7, 8, 9, 11 and 12, and for the information obtained at their printworks.

I am grateful to Nova Knit for information obtained from them and for permission to use the photographs of knitting machines.

To C. A. Crowther of A. E. Aspinall Ltd. and Eric C. Howarth of Bradford Technical College for technical information supplied by them.

Preface

Textiles play an important part in everyday life, both in household fabrics and in garment form. There are many courses of study and occupations in which a simplified knowledge of woven and printed textiles is required. This book has been written to assist those who wish to know about textiles, without going into the technology of the subject in great detail. It may also induce the reader to study the subject further, when the opportunity occurs.

It is difficult to explain some aspects of textiles in a simplified form, as it is a large and complicated subject. The ideal method of producing a book of this type would be to include samples of the different fabrics. As this is an impossibility, any reader will find it of great benefit to compile a notebook of cloth samples, to be used in conjunction with the text. A visit to a weaving mill or print-works, if at all possible, will help in the understanding of the methods employed to produce textiles. Some museums have sections on weaving and printing and these can also assist the student.

With the present rapid production of new fibres, those given in the book are as up to date as is possible. I have omitted tables on fibre structure and kept chemical data to the minimum to avoid confusion. In the sections dealing with the different types of looms, I have tried to classify, as far as possible, the fabrics produced on them.

This book is the outcome of a series of lectures that I have given to students, and it should provide others with the information required. Whereas in a lecture it is possible to explain certain points raised by students, in a textbook one has to anticipate the readers' questions and try to provide the answers. I hope the reader will forgive any queries that are not satisfactorily explained and will find something of interest and benefit in the book.   E.G.

Note on Metrication

Although metrication is now general in the retail trade, many manufacturers still use imperial terminology, or a mixture of the two styles. While, therefore, some metric expressions have been introduced in the present (1978) edition, no attempt has been made to metricate throughout.

CHAPTER 1

Natural Yarns: Growth and Manufacture of Vegetable Fibres

Fibres; Union Cloths; Staple and Filament; Natural Yarns; Cellulosic; Proteinic; Cotton; Linen; Jute; Hemp; Ramie; Coir; Sisal; Kapok; Paper Yarns;

INTRODUCTION

'Textiles' refers to all fabrics, whether in the piece or in garment form. Fabrics are either woven or knitted and each one has its own characteristics. These depend upon the kind of raw materials used, the class of yarns, the structure of the cloth and the addition of any decoration produced by printing or dyeing.

A textile must fulfil the requirements of everyday use, therefore it must be functional. Textiles are used for many purposes and these are primarily wearing apparel, household fabrics, furnishings, accessories and industrial materials. Every fabric differs in the qualities it must possess. These include the following:

Flexibility
Strength
Good draping
Retention of shape and size
Hard wearing properties
Softness
Porousness
Absorbency
Dye fastness to light and air conditions
Dye fastness to washing
Easy laundering
Easy dry cleaning

Textiles are purchased for their appearance, serviceability, and durability. The quality of the fabric is most important, and this depends on the kind of raw material used and the kind of manufacturing processes through which it has passed.

In order to understand the different qualities and acquire the ability to recognise various fabrics, a knowledge of the production methods is essential, combined with examining and handling the materials themselves.

FIBRES

The term *yarns* refers to all those threads used in weaving, which interlace to form a cloth or fabric and are manufactured from natural or man-made fibres. The natural are obtained from animals and vegetables and the man-made or synthetic are constructed chemically.

These two groups of fibres can be combined during the spinning processes, by the addition of a percentage of one group to the other, and spun as one yarn. This blending of fibres allows a wide range of yarns to be produced, each one having its own special qualities and uses. Fabrics are also woven by mixing natural and man-made yarns, e.g. the warp of natural and the weft of man-made or vice-versa.

These combinations of the two groups either by blending or mixing may reduce the cost of yarn, strengthen the fabric, lessen the weight, or benefit the yarn or fabric in other ways.

UNION CLOTHS

Originally woven with a linen warp and woollen weft, Union cloths are now produced in a variety of yarns, such as all natural, with a natural warp and man-made weft or vice-versa.

STAPLE AND FILAMENT

Natural fibres, with the exception of silk, are referred to as staple length fibres. The staple length refers to the limit of fibre growth. Wool is an example of this, as the staple length is taken from the root to the tip of the fibre and this varies according to the breed of sheep. A filament is a continuous single fibre and the filament length in man-made yarns may be several miles long. This filament can be cut into short lengths and spun as staple fibres.

NATURAL YARNS

The title natural is applied to all those yarns manufactured without the aid of chemicals, from animal hairs, parts of plants or vegetables and from the cocoon in the case of silk. The fibres from these groups have been used for thousands of years to produce yarn.

Textiles made from linen thread have been found in ancient Egyptian tombs, and the first cotton cloth is believed to have been woven in India. The Chinese were producing silk before the birth of Christ, and the actual process of obtaining this yarn was a closely guarded secret for many years. All these yarns were developed from man's early experiments in using the natural sources of fibres around him.

The range of fibres is limited, as only a certain number of animal and vegetable fibres are suitable, either economically or structurally, for producing yarns. The main factors governing this are flexibility, fineness, and a high ratio of length to thickness. Fibres are classed according to their molecular structure, the chain-like arrangements of molecules of great weight built up of many small units. The physical properties of a fibre depend on the chemical structure. Each fibre has its own characteristic appearance when it is analysed under a microscope, both longitudinally and in cross-section. This applies to both natural and man-made fibres.

Cotton has a corkscrew-like appearance and an oval cross section. A chemical process, called mercerising, straightens the corkscrew and makes the cross-sections more circular.

The ends of the single bast fibres are usually tapered to a point. Hair has a scale surface, and the size and shape of these scales varies from fibre to fibre. Silk is usually circular or triangular in cross-section, and longitudinally the fibres will appear cylindrical. From these different fibre structures and chemical properties the various yarns are produced, each one having its own special qualities and uses.

Natural fibres are divided into two classes, cellulosic and proteinic, which can then be subdivided in the following order.

CELLULOSIC Seed—Cotton
 Bast—Linen, Jute, Ramie, Hemp
 Leaf—Sisal
 Fruit—Coir
PROTEINIC Hair—Wool, Camel, Mohair, Worsted, Cashmere, Angora, Vicuna, Guanaco, Alpaca, Llama, etc.
 Filament—Silk

COTTON

Cotton is classified as a seed hair and has been the most widely used textile fibre. The cotton grows on bushes, three to four feet high, and requires from six to seven months of hot weather, plenty of sunshine and appreciable amounts of moisture. There are many varieties of cotton, all producing fibres of different qualities, varying in colour and staple length.

Sea Island cotton is very fine and has an average staple length of 1.6 to 2 in. (4 to 5 cm). Egyptian cotton is the most highly lustrous, with an average staple length of 1.4 in. (3.5 cm). The colour varies from pale cream to light brown. Indian cotton has the shortest staple length, usually not exceeding 1 inch, and can be a greyish brown in colour, and Coconada cotton is light brown.

The blossom appears and when it falls, the boll or seedcase begins its growth. The cotton fibres grow from the seeds, and when it is ripe the seed case splits open, exposing the fluffy white cotton. The bolls are usually picked by hand, or gathered mechanically. To separate the fibres from the seed, the cotton is subjected to a process of ginning in specially constructed gins. These machines have rotating brushes or fluted rollers to remove the fibres from the seeds, and the latter are collected for the manufacture of cotton seed oil. The fibres are packed into bales and sent to a spinning mill.

At the mill, the cotton is opened to remove the impurities, such as sand and grit. There are various kinds of openers, but they all consist of a beating mechanism which loosens the cotton. The opened fibres are blown against a perforated drum and the grit and sand are removed. The shorter fibres also pass through the drum, leaving the longer fibres to be rolled into a lap (a sheet of fibres). Several laps are passed through a scutching machine, and are converted into a fleecy mass of fibres ready for carding.

The carding machine has rollers covered with small pins or hooks, projecting from the surface. These pins pull the cotton fibres parallel to one another and, at the same time, separate them. The fibres are collected into a round sliver or rope and given a slight twist.

For high quality yarns the slivers are combed to remove all the short fibres. In the combing process, the slivers are pressed on to a series of rotating pins, and the fibres are pulled apart. The longer fibres grip more firmly and are then pulled away by drawing-off rollers, and the short fibres are lifted separately.

The slivers from either the carding or combing machine are drawn or drafted to reduce the thickness. They are drawn out through successive pairs of rubber rollers, each set rotating faster than the previous ones. The drawing process is continued further by subjecting the slivers to other drafting rollers. This reduces the slivers to rovings, which are fine enough to spin into yarn. Cotton rovings are usually spun on the ring and mule spinning machines. Cotton condensor is a yarn spun from waste cotton.

Diagram 1. Cotton Carding Engine. A = lap B = taker in roller C = cylinder D = doffer E = flats F = eyelet G = rollers H = sliver container J = sliver.
The lap (A) travels over the cylinder (C) after being taken in by the roller (B). The flats (E) a band of toothed cards move against the cylinder and the fibres are aligned by this action. The carded web is taken by the doffer (D) to the eyelet (F) and formed into a sliver (J) and drawn through rollers (G) into a sliver container (H).

Cotton contains carbon, hydrogen and oxygen and its strength increases approximately 25 per cent when it is wet. When cotton is treated with caustic soda, it becomes mercerised cotton. The caustic soda does not combine chemically with the fibre, but causes a physical change in the molecular structure of the fibre. This in turn changes its physical appearance and gives it a sheen.

LINEN

The flax fibres used to make linen fabrics are obtained from the inner bark or bast fibres of the flax plant. The plants have a straight stalk, with blue flowers, and grow to a height of 90 or 120 cm. White flowering varieties are more common on the Continent.

After flowering and before the seeds are fully ripe the plants are pulled whole from the ground by hand. Flax is pulled and not cut, as cutting not only loses fibre, but reduces the quality. In large fields the pulling may be done by machine. The seed pods are removed by passing the plants through an iron comb, a process known as rippling. The seeds are crushed to extract linseed oil.

Retting follows, and is a forced decomposition of the bark and core of the stalk in order to free the fibres. This is done by immersing the flax in water containing bacteria for a period of two to four weeks. Tests are made to ensure that over retting is not taking place, or loss of strength and colour may result. Steam retting is an alternative method and dilute sulphuric acid is added to the water.

When retting is completed, the stalks are washed in clean water and laid out to dry in the sun. Drying must be thorough or the retting action would continue. The flax bundles are then stored for a few weeks and sent to the scutch mill, where the loosened fibre is to be separated from the rest of the plant.

The plants are fed into a scutching machine, where they are beaten by a series of wooden blades. This removes a large part of the woody substance and bark and some of the short fibres. The long fibres are known as 'line' and the shorter fibres called 'tow'.

The flax then passes to the hackling machine, which combs out the fibres by using a series of pins, first coarse widely separated pins, then progressively finer pins. Hackling removes all the impurities, throws out the tow and leaves the long combed fibres fairly parallel. The long fibres are drawn into a sliver and passed through a series of drawing frames. The final drawing takes place on the roving frame and this imparts a slight twist to the sliver in order to strengthen it for the spinning frame.

Flax may be 'wet spun' by passing the roving through a trough of hot water, which softens it enough to be pulled out to the required fineness. The two most used methods of spinning linen are on the flyer or ring spinning machines. Linen yarn is used for a variety of fabrics, including dress, furnishings and household textiles.

JUTE

The jute plants grow to a height of 150 to 360 cm and in a normal season reach maturity in three and a half to four months. The plants are cut down by hand and subjected to retting for a period varying from eight to thirty days. During retting they are examined daily, and when the layers separate from the core easily, the retting process ceases. The stalks are stripped and dried, then sorted for quality of fibre. The jute is then sent to mills where it goes through similar processes to flax, finally emerging as a sliver. The slivers are then drawn on two frames and with a final operation to form the roving, the jute is wound on to a bobbin ready for spinning.

After spinning the jute is ready for weaving or dyeing. It cannot be bleached white as it disintegrates in strong bleaches and is most often used in its natural colour and spun into coarse yarns. The principal uses of jute are in the upholstery, linoleum and carpet trades.

HEMP

Some varieties of hemp are difficult to distinguish from flax, and when properly processed and spun, it is a good substitute for flax in larger size yarns. Hemp has never been able to compete strongly with flax as it lacks the fibre fineness. The production and manufacture of hemp is very similar to that of flax. The plants grow to a height of from four to eight feet with small yellowish-green flowers. The matured stems are usually hollow and the bark layer very fibrous throughout the whole length of the stem.

NATURAL YARNS: GROWTH AND MANUFACTURE OF VEGETABLE FIBRES

RAMIE

The plants grow from root cuttings to a height of 90 to 250 cm in a hot rainy climate, and are harvested by cutting, after which a new growth begins.

The ramie bundles are passed through a series of rollers in a decorticating machine. All the stalks are stripped and the woody portion removed. The fibre is then degummed; as much as 35 per cent of the decorticated fibre may be gum. All but 4 per cent of the gum is removed from the fibre in a mild chemical bath. The fibre length may range from 15 to 20 cm and is usually cut into desired staple lengths before spinning.

Ramie is a strong natural fibre and its strength increases when it is wet. Similar to linen, but pure white in colour, it has a silk-like lustre and excellent resistance to rotting, mildew, etc. Ramie has a tendency to break if folded repeatedly in the same place. In staple form it is blended with nylon, cotton, viscose and mohair and used for upholstery.

COIR

This fibre is taken from the husks of coconut while still green. It is softened by retting, rubbed to remove the tissue and dried. As it is too coarse for spinning into textile yarns it is used for mattings and upholstery stuffing.

SISAL

This fibre comes from the leaves of the Agave Sisalana tree, which grows mainly in Central America, Africa and the West Indies. The fleshy part of the leaf is removed by squeezing in machines and with a simultaneous washing the fibres are left. A further washing and drying takes place, leaving the fibres a creamy white colour. Sisal fibre is used for matting and twines.

KAPOK

Kapok is a seed fibre obtained from the kapok tree, and

grows in Java, Indonesia and the southern parts of Asia. The fibre is round, smooth and light, with a high lustre. It is used as a filling in upholstery and for cushions, mattresses and life saving apparatus. Kapok is often combined with waste cotton and rayon and spun into yarns for woven textiles.

PAPER YARNS

Paper yarns are manufactured as paper in the usual manner, but the fibres are placed lengthways and not in a random array, as they are in paper production. The sheets of paper are cut into strips, usually not less than 12 mm wide, and are twisted into yarn. The paper is moistened with water, before twisting takes place. To impart a stiffness or colour to the yarn, a stiffening paste or dye can be combined with the moistening process.

Paper yarns are strong, but do not stand up well to wetting treatments. They are principally used for carpet backings, and industrial fabrics.

CHAPTER 2

Natural Yarns: Source and Manufacture of Animal Fibres

Wool; Classification of Breeds; Re-Manufactured wool; Worsted; Mohair; Camel hair; Angora; Cashmere; Vicuna; Guanaco; Alpaca; Llama; Silk

WOOL

The fleece of the sheep is obtained by shearing once or twice a year, in such a manner that it remains in one piece. Wool taken from a dead sheep is referred to as pulled wool and is of inferior quality as the lime water used to remove it softens the roots. Each breed of sheep produces a different type of wool and the staple length may vary from 4 to 25 cm. The quality of wool depends on the length and diameter of the fibres, and other physical properties such as strength, elasticity, shrinkage, colour, lustre and crimpiness. Quality is also determined by the part of the body from which the wool comes, the large areas such as shoulder and back producing a better quality than the tail or legs. The fleece is greasy, as wool grease is exuded on to the wool where it forms a protective covering for the fibres against the weather.

The fleece is first graded, then sorted into various qualities. This is done by hand as it demands considerable skill and experience, depending on visual examination and feel. The various wools are designated 'combing' or 'carding' wools according to whether they are best suited for use in the manufacture of worsted or woollen yarns. Wool contains impurities such as dried grass, thistles, dried earth and the natural impurities of wool fat or wool sweat and suint (potassium salts). These impurities may make up

50 per cent of the weight of raw wool and have to be removed before the carding and combing operations, preliminary to spinning. The recovered fatty matter is lanolin which is extracted and purified.

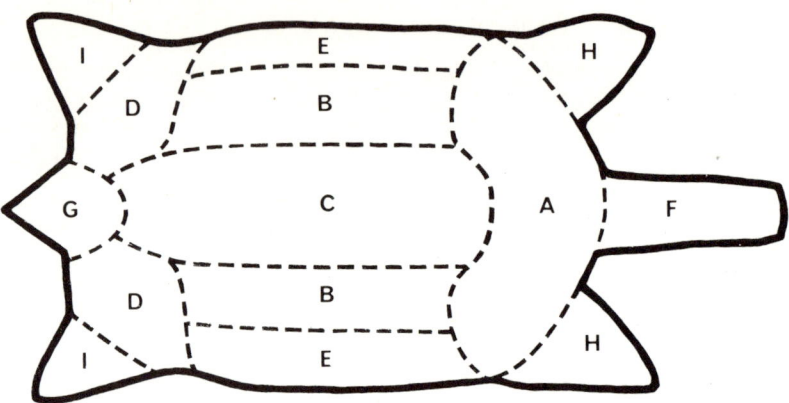

Diagram 2. Fleece. A = shoulders—best quality B = sides—not quite so fine or regular C = back—less close and strong than shoulders D = haunches—coarse and long E = belly—fine, short, sometimes felted F = neck—fine, short and often irregular G = tail—very coarse and often kempy H = forelegs—short, irregular and poor I = hind legs—coarse, hairy, strong.

The fleece is opened up and scoured or washed by passing it along a machine through consecutive tanks containing detergents and a mild alkali in the first tank and finally rinsing water in the last one. Incorrect washing may lead to deterioration of the fibres. It is essential that all vegetable matter should be removed, and in order to achieve this, the wool is carbonised by wetting it with hydrochloric acid followed by heating. This burns all the vegetable matter and leaves the wool unaffected so that after drying the unwanted substance can be dusted out. The woven cloth can also be carbonised. Wool can be dyed in the loose state (stock dyeing), a process which occurs before spinning.

NATURAL YARNS: SOURCE AND MANUFACTURE OF ANIMAL FIBRES

The dyed and undyed fibres, or fibres of different qualities, are opened out on the Teaser and thoroughly blended. Oil is applied to assist in the spinning. Wool can also be blended with different fibres to form a union yarn.

The fibres are carded in a carding machine and converted into slivers of a uniform quality and weight. These are cut into narrow ribbons and rolled into soft twistless slivers to be wound on to bobbins. Before the spinning process, the slivers are condensed into rovings by stretching or drawing them on a series of rollers. The rovings are usually spun on the mule spinning frame.

CLASSIFICATION OF BREEDS

SHROPSHIRE: This wool is used in the Welsh flannel trade. It is very soft and white in colour.

SOUTHDOWN: The fleece is very short in staple length, close and crimped. Scarves and hosiery are produced from the yarn.

CHEVIOT: Produces good quality wool from a 4-inch staple and is dense and soft. It is used for tweeds and fine woollen dress fabrics.

SUFFOLK WOOL: This is less fine than Southdown and has a lot of dark hair. Principally employed in the tweed and hosiery trades.

LINCOLN AND LEICESTER: Noted for strength and lustre. Used for dress fabrics, linings and in some cases serges and broad-cloths for heavy wear.

ROMNEY MARSH: This wool is a bright and typical demi-lustre wool.

BLACK-FACED HIGHLAND: Wool is used for the carpet trade.

SHETLAND: Native to Orkney and Shetland and produces a wool for tweeds and knitwear.

RE-MANUFACTURED WOOL

All kinds of woollen materials, both old and new, are used to reproduce woollen yarn. These materials include cast-off woollen and worsted clothing, knitwear, pattern books, blankets, etc. After sorting, they are ground up into fibrous material and re-processed.

Shoddy, a term meaning 'shed' or 'cast off' is made from fibres of a good length obtained from knitwear and tweed. These reprocessed yarns are employed in the manufacture of woollen suitings, coat materials, blankets, etc.

Mungo is manufactured from felted woollens and worsted cloths. The fibres are very short and are blended with new wool or cotton and spun into coarse weft yarns.

Extract is the name given to a fibre obtained from a union fabric. This is generally wool from a wool and cotton union. The cotton is destroyed by sulphuric acid or by other chemical treatments and leaves the short-fibred wool. This is blended with other fibres to produce a yarn.

WORSTED

In the production of worsteds, the chief object is to produce a smooth round yarn of parallel fibres. Woollen and worsted yarns may be spun from the same fleece. The difference between the two lies in the preparation prior to spinning. In worsted yarn the fibres are straight and parallel, whereas in woollen yarn the fibres are crossing and re-crossing each other.

The fleeces used for worsted yarn are the finest merinos, which produce the best quality of yarn. Other worsted yarns are manufactured from all the other types of crossbreds with a suitable staple length.

The wool fibres are opened out and carded to separate the short and long fibres. The shorter ones are designated for wool and the others for the manufacture of worsted yarn. The longest fibres are converted into slivers, and washed in a scouring liquid.

The slivers are passed between heated cylinders where they are stretched and dried, thus reducing the crimp of the fibres so that they can be more easily aligned.

Oil is applied and they are straightened out in a gilling machine. The sliver passes between a pair of fluted rollers in the gill box and through a series of pins fixed on moving bars (pinned fallers). It leaves the gill box by means of a second pair of rollers rotating at a greater speed than the first set of rollers. Gilling reduces the sliver whilst at the same time aligning the fibres in a parallel manner. Several gilling operations are carried out to prepare the slivers for combing.

Diagram 3. Worsted Carding Engine. A = swift B = wire covered rollers C = fancy D = doffer E = stripping rollers F = working rollers G = sliver rollers H = comb J = sliver container.

The lap is taken in by rollers (B) and travels over the swift or cylinder (A) and is carded by the action of the working rollers (F). The lap is removed by the stripping rollers (E) and cleared by the doffer (D) and comb (H) and emerges through the rollers (G) as a sliver.

As combing takes place the long fibres (tops) are separated from the short ones (noils) by the action of the mechanical combs and brushes. A final gilling completes the preparation and the sliver leaving the finisher is given a twist by passing it through a flyer.

The spinning of worsted yarns is carried out mainly on the ring spinner (see page 39). For a smooth fabric it is gassed, a process that burns off all the loose fibres. Botany Quality is the term used for a fine worsted yarn.

MOHAIR

Mohair is the fibre from the Angora goat and has a staple length of from 10 to 15 cm for six months of growth. The raw fibres are of a slightly yellowish or greyish tint and are white after washing. Mohair is a strong resilient fibre, with very little crimp and a good affinity for dyes. It is blended with wool, silk, cotton and rayon and is also used in combination with these fibres. The medium quality fibres are spun for use in upholstery and rugs.

CAMEL HAIR

The camel hair used in textiles comes from the Bactrian or two-humped camel found in Central Asia, but only a small quantity comes to England. It frequently contains vegetable matter and to remove this the hairs are carbonised. The noils or undercoat are the hairs used for textiles and these are made up into blankets, coats, dressing gowns, etc. Pure camel hair material is expensive and a small percentage of wool is blended with the fibres to reduce the cost of some fabrics. This blended yarn is also used for knitted garments.

ANGORA

Angora is the hair from the Angora rabbit, which was originally raised in France and North Africa. In France, the fibres are sent to a central agency where they are sorted and graded. The fibres are combed to obtain the longest staple length possible and each rabbit produces only a few ounces of hair. As the rabbits are difficult to raise the total quantity of angora is very limited and expensive. Angora is fairly long in staple length, fine, fluffy, soft and slippery, requiring special processing to spin it properly. It is pure white in colour, and used for knitted garments, or mixed with a percentage of wool in the manufacture of dress fabrics.

CASHMERE

This fibre comes from a goat smaller in size than the Angora goat, and takes its name from the province of Kashmir in Northern India. The hair is combed by hand from the animal during the moulting season. Two kinds of fibre are obtained, coarse long hairs, which are exported as goat's hair, and very fine, soft fibres that are used for the finest fabrics. The natural colours of the fibres vary from white, grey, to brownish grey.

The fibres are put through a highly specialised process of mechanical combing to remove all the coarse hair. The yield of this finished product goes into the manufacturing of fine cashmere fabrics, and is usually not more than 220 g per goat. Pure cashmere fabrics and knitted garments are expensive and a small percentage of wool is often blended with the cashmere fibres, thus reducing the cost.

VICUNA

The vicuna is a wild animal of the South American branch of the camel family, and stands about 90 cm high. It can only be domesticated if captured very young, and this is difficult because of the inaccessible habitat in which it lives on the highest peaks of the Andes.

Vicuna is the softest, finest, rarest of all textile fibres as well as the most expensive one. The fibre is short, very lustrous and a light cinnamon in colour.

GUANACO

The guanaco is larger than the vicuna, with downy hair, the difference in fibre being in fineness and the colour which is a little deeper and of a reddish brown. The very young guanaco, known as guanaquito, are usually taken between seven and ten days after birth for their pelts and are a pale sand colour. Guanaco fibres are manufactured as worsted and used for fine woollen fabrics, but are rather expensive in comparison with wool and worsted cloths. The fibres are blended with wool fibres for some materials.

ALPACA

The alpaca is a small animal about 105 cm high, with a long fleece reaching almost to the ground and a staple length varying from 20 to 30 cm. The animals are clipped every second year and yield 2 to 4 kg of fibre per animal. The natural colours of alpaca are white, light fawn, light brown, dark brown, grey, black and piebald, and the fibres are noted for their softness, fineness and lustre. Alpaca is an expensive fibre used alone and can be blended with other hair fibres.

LLAMA

The llama is a South American animal with a long fleece of fine and coarse hair. The fibres are difficult to separate and require careful preparation for spinning. The natural colours are the same as those of the alpaca, and the llama fibres may be spun or blended with wool for use in the manufacture of dress and coat materials.

SILK

There are many varieties of silkworm, as the caterpillar is called, and Bombyx Mori is the species most commonly bred in Europe. It makes the cocoon by squirting a fluid from two glands while two other glands secrete gum (sericin). The fluid and gum coagulate on contact with the air and form a silk thread. The cocoon, consisting of about 3.5 km of continuous solidified filament firmly gummed together, is about the size of a peanut shell.

The chrysalis is killed by subjecting it to heat and the gum is then softened by boiling water. The cocoons are placed in a vessel and the loosened ends of the filaments adhere to the bristles of a revolving brush. As the brush is raised the cocoons begin to unwind. One filament is too

NATURAL YARNS: SOURCE AND MANUFACTURE OF ANIMAL FIBRES 19

fine for practical purposes, so from 3 to 8 cocoons are unwound and reeled together, to form one single yarn. After the filaments have been made into yarn the gum is removed. Silk is strong, elastic and lustrous, with a good affinity for dye.

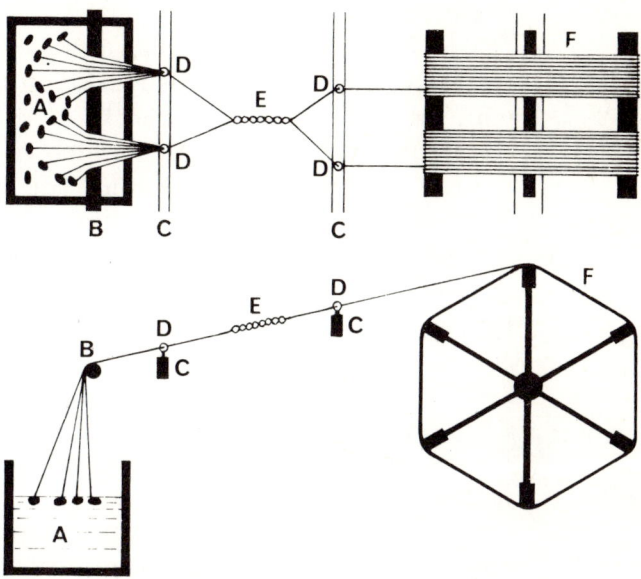

Diagram 4. Silk Reeling. A = cocoons in hot water vessel B = glass rod C = bars D = eyelet holes E = crossed filaments F = skeiner. The filament ends are led from the cocoons in a vessel of hot water (A) over a glass rod (B) and through the eyelet holes (D) of bars (C). The filaments cross at (E) and are wound into a skein on the skeiner (F).

RAW SILK: An untwisted silk thread in skein or hank form which is produced by reeling together the filaments from several cocoons. The sizes of raw silk varies from 8/10 deniers to about 20/22 deniers.

FILATURES OR STEAM FILATURES: This term is applied to raw silk reeled by machinery.

CHINA SILK: Raw silk reeled in northern China. The filatures are fairly uniform in size, of good quality and white in colour.

DOUPPION SILK: Raw silk reeled from double cocoons, which are united by worms having spun their cocoons together. The cocoons are reeled from alternately, and give an uneven thread. Thick yarns and cheap fabrics are produced from this silk.

TRAM SILK: A weft yarn consisting of two or more untwisted or slightly twisted singles, run together and twisted with two or three turns per inch.

ORGANZINE SILK: A warp yarn composed of two or more hard twisted singles, wound together and given a hard twist. This yarn is not so lustrous as tram.

TUSSAH SILK: This is produced from the cocoons spun by wild or uncultivated silkworms, and is a thicker and less lustrous fibre than from the cultivated worm. The wild silk is brown in colour.

SPUN SILK: A yarn made from waste silk; the term includes every kind of raw silk that cannot be reeled, such as defective or pierced cocoons and the waste silk accumulated in the various processes. Long raw fibres are cut to a maximum length of about 30 cm and are spun as for worsted, and short lengths are spun like cotton.

CHAPTER 3

Man-Made Yarns: Origin and Production

Synthetic: Cellulosic; Proteinic; Alginate; Mineral. Polyamide Fibres: Nylon; Bri-Nylon; Celon; Perlon; Enkalon; Fluflon; Miralon. Polyester Fibres: Terylene; Crimplene; Dacron; Fluflene; Miralene. Acrylic Fibres: Acrilan; Courtelle; Orlon; Dralon. Modacrylic Fibres: Teklan; Vinyon. Polythene Fibres: Courlene. Polyurethane Fibres: Spanzelle. Polypropylene Fibres: Ulstron. Cellulosic Fibres: Viscose Rayon; Vincel; Evlan; Durafil; Sarille; Fibro; Rayolanda. Cellulose Ester Fibres: Acetate Rayon; Dicel; Tricel; Tricelon; Celafibre; Cuprammonium Rayon. Proteinic Fibres: Fibrolane; Ardil; Vicara; Alginate. Mineral Fibres: Glass; Asbestos.

Man-made fibres are produced by the chemical treatment of certain raw materials, such as cotton linters, wood pulp, extracts of petroleum and the by-products of coal. These materials are converted into a spinning fluid by the addition of chemicals, and forced at pressure through the holes of a spinneret. The spinneret used today is copied from the spinner in the head of the silkworm. It is a disc with a series of small holes, and as the molten material is forced through, it forms continuous strands of fibre. These are used in filament form as yarn, or can be cut into staple lengths and spun. A greater number of filaments are extruded for the staple length fibre than for filament yarn. The spinneret is comparable to the head of a watering can or shower in miniature, the water being replaced by the fluid, which emerges in the same manner.

It is thought that the first man-made yarns, in the form of a gelatinous substance, were made in ancient China. The first use of a man-made spinneret seems to have been carried out by Louis Schwabe in 1842, using molten glass to produce a thread. Sir Joseph Swan, the electrical pioneer, patented nitro-cellulose filaments, which his daughters used for embroidery threads in 1883. Count Hilaire de Chardonnet also worked on this type of filament and was the first man to produce any type of artificial fibre on a commercial scale.

From these beginnings other chemists and scientists worked on the production of man-made fibres. Factories were built to produce the new yarns, and the research with different materials is still carried on, with new yarns being produced all the time. The main features of all synthetic fibres is their plasticity and the long chain molecules in the fibre, which are mobile enough to allow solubility in solvents and melting at a high temperature. Man-made fibres can be accurately controlled and modified during manufacture, and are divided into five classes in the following order:

SYNTHETIC   Nylon, Terylene, Orlon, Acrilan, Dacron, Courtelle, Courlene, Ulstron, Teklan, Vinyon

CELLULOSIC   Viscose, Acetate, Cuprammonium (Rayons), Durafil, Sarille, Evlan, Duracol, Vincel, Dicel, Tricel

PROTEINIC   Casein—Fibrolane
Peanut—Ardil
Maize—Vicara

ALGINATE   Seaweed

MINERAL   Glass
Asbestos

POLYAMIDE FIBRES

NYLON

Nylon is produced entirely from mineral sources, the raw materials are benzine from coal, oxygen and nitrogen from the air and hydrogen from water. A nylon salt is formed and converted into a polymer. The nylon polymer, resembling plastic, is cut into the form of small chips. These are fed into a hopper and from there into a melting unit. The molten nylon is pumped through tiny holes in a spinneret, and as the filaments emerge they solidify on contact with the air and are gathered together into yarn. The yarn is stretched or 'cold drawn' between a series of rollers. Drawing aligns the long chain-like molecules, places them parallel to one another and brings them closer together. The fibre diameter is reduced by drawing and the nylon becomes very strong, tough and elastic, with a translucency and lustre.

Nylon staple is made by crimping continuous nylon filament tow, and by cutting it into short uniform lengths, it can be reprocessed on cotton, linen or wool spinning machines. Under a microscope, nylon resembles cuprammonium rayon and dacron. Nylonising is a finishing process to increase the natural water repellency, so that moisture spreads rapidly and evaporation is speeded up. This finish gives quick drying fabrics and also increases the softness, pliability and warmth.

A heat set can be applied to the woven fabric or to the yarn. This bonds the molecular chains together and gives a shape and size that can only be altered by exposing it to a higher temperature. Nylon is resistant to most chemicals, but is damaged by strong oxidising bleaches and concentrated acids; therefore a mild bleach is used for bleaching the yarn or fabric. It is easy to dye and has an excellent fastness, with tones ranging from pale pastels to deep shades.

Nylon is not resistant to sunlight, but bright nylon is more resistant than delustred nylon as less light is absorbed: it is also unaffected by salt water, moths, mildew, and other organisms. Nylon melts, when exposed to a flame, but does not burn, and is regarded as one of the safest of all textile fibres as far as inflammability is concerned. It is used for children's clothing, men's and women's wear and household fabrics, either alone or blended with other fibres.

BRI-NYLON

A nylon yarn mainly used in the manufacture of carpets, lingerie, nightwear, and lightweight knitwear. It wears well and the surface can be brushed to form a short pile.

CELON

A nylon yarn produced in coarser denier filament. It is used in stretch fabrics, quilted materials, upholstery, rainwear and other garments. Celon is produced in white and does not discolour after repeated washing. It also has excellent flame resistant properties.

PERLON

A fibre with properties and uses similar to nylon. It is blended with rayon, cotton or wool fibres and improves the wearing qualities of these other fibres. Perlon is principally used in the clothing industry.

ENKALON

A nylon yarn with primary uses in upholstery, loose covers, and carpets. Filament yarn is exploited fully in warp and weft knitted cloths which have good extension and recovery for upholstery. Enkalon has good resistance to wear and fastness against light, rubbing and washing. It is a little difficult to achieve colour fastness in some brighter shades. Various colour combinations can be produced by spun dyed yarns.

FLUFLON

This is a modified nylon yarn, produced for outer garments, shirtings, knitted fabrics, carpets and furnishings.

MIRALON

A crinkle or bouclé type yarn in nylon.

POLYESTER FIBRES

TERYLENE

Terylene is produced from petroleum and its by-products, mixed with various chemicals, including Ethylene Glycol and Terephthalic acid from which it gets its name. A polymer is formed through a series of chemical processes, and extruded on to a casting wheel, where it solidifies. The ivory-coloured plastic polymer is cut into small chips and from this point Terylene is produced in two distinct forms.

In the production of filament yarn the polymer chips are melted at a high temperature. The molten fluid is extruded through the spinneret, and as the filaments appear, they solidify. The yarn is drawn or stretched to several times its original length, and wound on to bobbins. The Terylene filament yarn may then be further processed according to its end use.

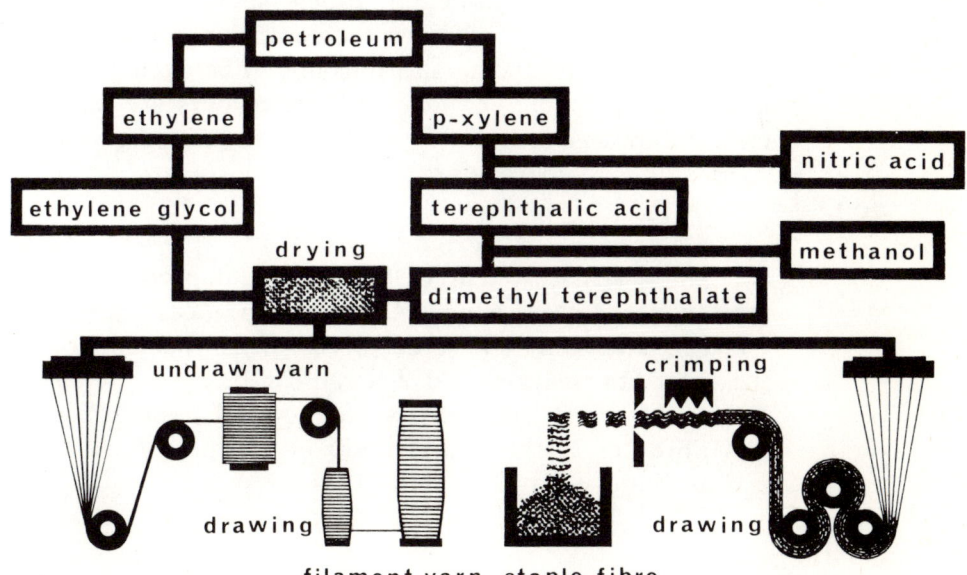

Diagram 5. The Manufacture of Terylene.

Staple fibre is manufactured by melting the polymer and extruding it through the spinnerets. Many more filaments are extruded together than for the filament yarn. These are brought together into a thick tow to be drawn, and crimped with an artificial wave which is then set by heating. The tow is cut into specified lengths suitable for spinning on the cotton, worsted, woollen or flax systems. Terylene staple fibre can be spun by itself, or blended with other fibres.

Terylene yarn is used for the manufacture of clothing and furnishing fabrics. It is fire resistant and melts when a flame is applied to it.

CRIMPLENE

A continuous filament yarn with less shrinkage than the normal Terylene fibre. Crimplene can be knitted much closer to the finished size and texture of a garment and therefore makes production much more simple. The properties of Crimplene allow the complete garments to be made up prior to dyeing and finishing. The dyeing process is shorter than for Terylene and the colours are fast and clear. Crimplene in jersey form is used for knitted garments, swim suits and sportswear.

DACRON

An American equivalent of Terylene and produced from the same sources. It is naturally white and does not require a bleach. In blends with other fibres, any bleach suitable to the other fibre can be used. Dacron can be dyed any colour and washing and light fastness are good. The main uses of this fibre are in the manufacture of clothing, furnishings and as fillings for pillows and quilts.

FLUFLENE

A modified Terylene yarn, produced for the manufacture of socks, stockings, underwear and outer garments.

MAN-MADE YARNS: ORIGIN AND PRODUCTION

MIRALENE

A crinkle or bouclé type of yarn based on Terylene.

ACRYLIC FIBRES

ACRILAN

Acrilan is an acrylic fibre produced mainly from acrylonitrile, which is a liquid derivative of oil refining and coal carbonisation processes. The acrylonitrile is converted into a polymer and dissolved in a solvent to form a spinning solution. This is filtered and the moisture content is reduced.

The spinning solution passes through metering pumps to the spinnerets and is extruded into a spin bath in the form of filaments. The tow bundles are passed over drying drums to reduce the moisture content. For staple fibre the tow bundles are crimped and have lubricants and anti-static agents applied. After crimping, the tow is cut into various staple lengths of from 2 to 15 cm.

Acrilan is used alone or blended with other fibres in the manufacture of all kinds of clothing, blankets, fabrics and carpets. It has good resistance to acids, mildew and is mothproof. Fabrics made from Acrilan are hardwearing, easy to clean and have a good resistance to abrasion.

COURTELLE

A true synthetic, produced from oil refining and coal carbonisation processes. The manufacture of the yarn is the same as for Acrilan. The fibre is produced in tow or staple form and used alone or in blends with man-made and natural fibres. Courtelle is used alone for a wide variety of garments including blankets and carpets. In blends with other fibres it is used for clothing and furnishing fabrics. The crease resistance and strength of Courtelle is good and it is unaffected by moths and mildew.

ORLON

Orlon is an American fibre made from acrylonitrile derived

from ethylene oxide and hydrocyanic acid and extruded through a spinneret to form Orlon filaments. The fibre is drawn to orientate the molecules and give it strength. In filament form it is stronger and has more sun and stretch resistance. An Orlon filament called Type 81 is semi-dull, off white, resembling the texture and lustre of silk.

In staple form two main types have been developed: Type 41 which is similar to filament Orlon, and Type 42 which is used in piece dyed fabrics (see later). It has excellent brilliance and fastness with some basic and acetate dyestuffs. Staple fibre is more resilient and more dyeable than filament fibre, with a special affinity for copper, which in turn has an affinity for certain dyestuffs. Orlon staple has a wider use blended with wool than in the worsted type of clothing fabrics. It blends well with cotton and nylon and gives a better fabric than a 100 per cent Orlon cloth.

With more sensitivity to heat than Dacron or nylon, Orlon melts and burns leaving a black plastic bead. It does not dissolve in the solvents used for other fibres. The main uses of Orlon are in knitwear, dress fabrics and furnishing fabrics.

DRALON

A synthetic derived from oil and coal and produced in West Germany. It is mothproof and has a good resistance to acids and alkalis and exposure to sunlight. Dralon absorbs hardly any moisture and is therefore quick drying and does not shrink or felt. This fibre is used for upholstery in either flat or pile woven fabrics and for net curtaining. It is also blended with natural fibres, but this reduces its quick drying and shrinkage resistance properties.

MODACRYLIC FIBRES

TEKLAN

A British fibre based on acrylonitrile and vinylidene

chloride derived from oil and coal. This yarn is flame-resistant, strong and hard wearing, but soft, warm and light. It has good resistance to sunlight, bacteria and most chemicals.

The main uses of Teklan are in woven and knitted dress materials and household textiles such as net curtains and furnishing fabrics, women's lingerie and children's nightwear.

VINYON

This fibre made from vinyl chloride and acrylonitrile is used for protective clothing and fur fabrics. It is flame-resistant and melts at a high melting point. Vinyon can be made into tea bags and other fabrics such as bathing suits, gloves, upholstery and millinery cloth. It shrinks at 65°C and cannot be used in garments or other fabrics that are ironed at ordinary temperatures.

POLYTHENE FIBRES

COURLENE

This yarn, produced from propylene gas, a by-product of oil, is fast and resistant to staining, acids and alkalis. It has high strength and toughness with uses in various fields such as deckchair fabrics, awnings and blinds, upholstery and dress fabrics.

POLYURETHANE FIBRES

SPANZELLE

A synthetic elastomeric fibre, which is produced as a multi-filament yarn. It is made in a range of deniers and has excellent stretch and recovery properties. Spanzelle has particular application in foundation garments, swimwear and for support stockings.

POLYPROPYLENE FIBRES

ULSTRON

This fibre is produced from by-products of oil and was first used for industrial ropes, cordage and fishing nets. It is very strong, and has a good resistance to abrasion and to most acids and alkalis. In staple form Ulstron is used for manufacturing blankets, and in filament and staple it is used for upholstery fabrics and carpets.

CELLULOSIC FIBRES

VISCOSE RAYON

The cellulose raw material is usually derived from wood pulp. The woods most favoured are the Norwegian or Canadian spruce, beech, pine and hemlock.

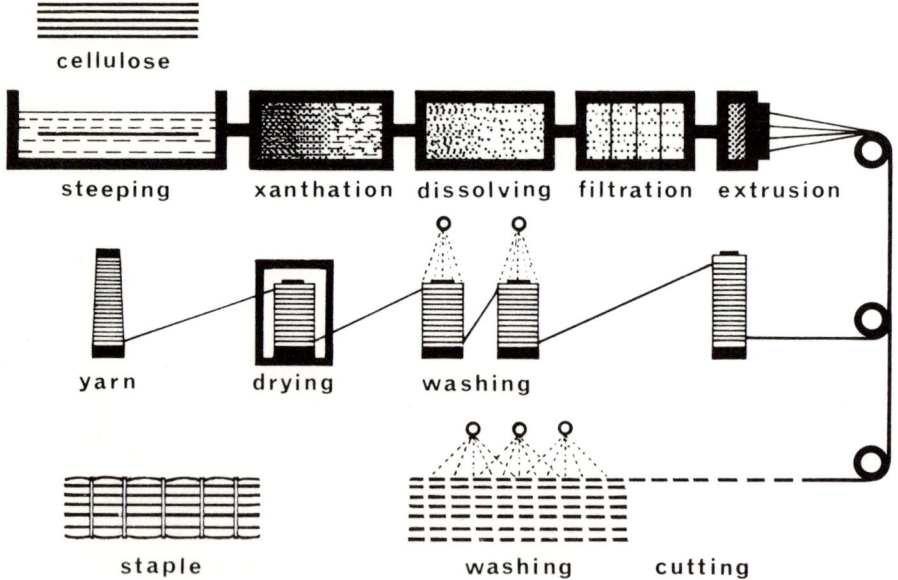

Diagram 6. The Manufacture of Viscose Rayon.

The logs of wood are cut up into small chips, less than 1 inch long and ½ inch wide, and treated with an acid solution. To form 'wood pulp', the solution is steamed for about 15 hours and then bleached. It passes through several rotary presses to remove the excess moisture, and is hardened into thin boards or sheets.

The parchment-coloured sheets of cellulose are steeped in caustic soda and ground until they look like white fluff or flock. This is left for a few days to allow action between the soda and cellulose, then liquid carbon disulphide is poured on to it. As the cellulose dissolves, a sticky yellow or orange substance is formed.

This treacly mixture is forced through the spinneret and emerges as threads, which are toughened by immersion in a bath of dilute sulphuric acid. Viscose rayon is produced as filament yarn or in staple form and used for women's, men's and children's wear, furnishing fabrics, carpets and other household textiles.

VINCEL

The physical properties of this yarn closely resemble cotton, although it belongs to the viscose fibres group. In woven fabrics and knitted garments, it is used on its own or blended with other man-made and natural fibres.

EVLAN

A modified rayon fibre developed for the carpet industry, which gives a good cover and resistance to crushing. It gives clarity of shade and reduces the degree of soiling that takes place in carpets. Evlan is frequently blended with nylon, wool and Courtelle, for heavy and light weight tufted Axminster and Wilton carpets.

DURAFIL

This fibre is developed from viscose rayon and gives an added toughness to many types of fabrics. It has high strength and abrasion resistance and, blended with other man-made or natural fibres, it improves the wearing qualities of the fabrics.

SARILLE

A fibre developed from viscose rayon and used to produce fabrics with wool-like characteristics, as it gives a warm, full handling texture. The main uses of Sarille are in dress fabrics, children's and men's wear, also household textiles, such as blankets and candlewick covers.

FIBRO

A viscose staple fibre, it can be spun alone or blended with wool, cotton, etc. The blending of Fibro with wool facilitates the spinning and enables a finer yarn to be spun, at the same time giving added strength to the yarn. The dyeing of the blended yarn or cloth is a costly process as solid shades are difficult to obtain with ordinary Fibro. The main uses of this fibre are in curtain and upholstery fabrics where it is used for weft or in blends with other fibres for warp and weft.

MATT FIBRO is a dull fibro, which has been delustred in either yarn or staple form.

SPUN DYED FIBRO is a staple fibre, and is spun by injecting pigments into the spinning solution. This gives a fibre with the colour built in.

RAYOLANDA

This is a development of Fibro, which has been given dyeing properties, so that dyestuffs used for wool and cotton may be applied to it.

Rayolanda is spun alone or blended with wool, cotton and other suitable fibres for a variety of clothing and furnishing fabrics. It has good resiliency and is warmer and softer than ordinary Fibro.

CELLULOSE ESTER FIBRES

ACETATE RAYON

A solution is made of acetic acid, acetic anhydride and a small amount of zinc chloride or sulphuric acid. Cotton linters (short unspinnable cotton hairs) are placed into this solution and left for several hours with the temperature slowly raised. This process is called 'acetylation', and after it has taken place a thick fluid resembling treacle is obtained. The fluid is allowed to ripen and the cellulose acetate is separated from the mixture by adding water. The acid is removed and what remains is cellulose acetate in solid form. This is dried and ground up until it becomes a flaky white substance. Acetone is added and the fluid after filtering is extruded through the spinneret to form acetate rayon yarn.

Acetate rayon has good dyeing properties in yarn and fabric form. Coloured dyestuffs are sometimes added to the spinning solution prior to extrusion. This rayon is absorbent, with a high resistance to moths and mildew. It has exceptionally good draping qualities and is used extensively in the manufacture of a wide range of fabrics, including men's, women's and children's wear and also furnishings. In staple form it is blended with a variety of fibres.

DICEL

A cellulose acetate fibre made from cotton linters or wood pulp, with the appearance and feel of silk. Dicel staple fibre is used for many industrial purposes. In dress fabrics it drapes well and can be made stainproof by a silicone finish. Dicel yarns have a high standard of colour fastness and the colour can be added to the solution before the filaments are spun.

TRICEL

The base of Tricel is either wood pulp or cotton linters. The cellulose is treated with acetic acid, and acetic anhydride obtained from petroleum and converted into cellulose triacetate. The cellulose triacetate is a flaky white material and by dissolving it in a solvent, a spinning solution is obtained. This is extruded through the spinneret and drawn out as filaments which are passed through a cylindrical tube and a current of warm air. The solvent is evaporated by the warm air and leaves the solid filaments to be twisted and wound on to a bobbin as continuous filament yarn.

Tricel is used for a wide variety of clothing and furnishing fabrics. In staple form it is blended with cotton, nylon, wool and viscose rayon. Filament yarn is woven and knitted into fabrics for dresses, blouses, lingerie, ties and coat linings.

TRICELON

A new yarn produced from Tricel and Celon, and woven or knitted for dress fabrics, lingerie, nightwear and men's shirtings.

CELAFIBRE

This yarn is an acetate cut staple fibre, which can be used alone or blended with wool for heavy fabrics and blankets. It is also blended with cotton and gives great possibilities of cross dyeing (see page 185). Celafibre used in woven fabrics gives crease resistant properties to the cloth.

CUPRAMMONIUM RAYON

The raw material is usually cotton linters, dissolved in copper sulphate combined with soda and ammonia. The mixture is continuously stirred and becomes a thick fluid which is filtered through very fine screens, and extruded through the spinneret. The elasticity of the filaments is considerably toughened by passing them through a coagulating liquor.

The separate filaments are drawn out, twisted together and stretched to form a single thread; this process is called stretch spinning. The yarn then passes through a setting bath consisting of hot water, and has acid applied at the final stage. Little of this yarn is made today.

PROTEINIC FIBRES

FIBROLANE

The basis of this yarn is regenerated protein fibre derived from casein, which is a by-product of milk. The casein is subjected to filtration processes and is mixed with caustic soda and water. A mixture is formed and undergoes de-aeration in a vacuum chamber. This solution is extruded through the spinnerets into a coagulating bath, and drawn off over rollers. The continuous filament is cut into short staple lengths, washed and dried.

Fibrolane is a resilient white fibre, which is absorbent, warm and soft, with good dyeing properties. It is only used in staple form and is blended with wool, rayons, nylon and other fibres. These blended yarns are employed in the manufacture of furnishing fabrics and carpets.

ARDIL

This fibre is a protein rayon produced from peanuts. It is cream coloured, crimpy and resilient and dyes like wool. In staple form it is light and soft and is best used in blends with wool, cotton or other rayons.

VICARA

A fibre produced with maize combined with chemicals. It was of limited use and production in the U.S.A. has ceased.

ALGINATE

A yarn produced from seaweed which is treated with acids and made into a spinning solution to be spun like viscose rayon. The material in yarn form or in the finishing of woven or knitted fabrics is treated in a bath containing metallic salts. The main use for this metallic alginate rayon is in furnishing fabrics.

MINERAL FIBRES

GLASS

The fibres are produced by subjecting small glass globules to electrical heat rising to 1,450°C. The molten glass is drawn off through tiny holes in a crucible on to spindles, which at the same time extend the filaments to a very fine degree. The finer filaments are less brittle and give better draping qualities to the woven fabric. Coloured yarn usually has the colouring matter applied before the drawing process.

ASBESTOS

Asbestos is produced from a mineral fibre, found as veins in a certain type of rock. The rocks are broken and the fibres are separated. The fibre is smooth and fine, and can be spun alone, or with a percentage of cotton. Asbestos yarns are woven into various fabrics, the main one being protective clothing. The longest staple fibres are spun into yarn and woven in other textiles.

CHAPTER 4

Spinning Methods

Flyer Spinning; Ring Spinning; Cap Spinning; Mule Spinning. Folded or plied Yarn; Twist of Yarn; Fancy or Novelty Yarns; Spiral Yarn; Gimp Yarn; Snarl Yarn; Knop Yarn; Cloud Yarn; Tuffle Yarn; Nep Yarn; Curl or Loop Yarn; Boucle; Grandelle Yarn; Marl Yarn; Nub Yarn; Flake Yarn; Chenille Yarn; Lurex. The size of Yarns; Simple Tests to Identify Yarns; Burning Tests; Caustic Soda Tests

The object of spinning is to draft or draw a roving into yarn of a specified fineness and to insert sufficient twist to bind the fibres together. The yarn must be evenly spun and strong enough to withstand all future strain imposed on it.

The earliest spinning of yarn was done with a spindle, a tapered stick of wood, notched at the top with a weight at the bottom. As the spindle revolved like a top the fibres were drawn out and twisted together, then wound on to the spindle and the process re-started. By this method it was possible to spin yarn whilst walking or moving about. When the spinning wheel was invented in the fourteenth century the amount of spun yarn was increased. The invention of Arkwright's Spinning Jenny enabled yarn to be spun commercially for use in the textile weaving mills.

All fibres have their own properties and characteristics and it is not possible to spin them all on the same machine; various spinning frames have been developed to deal with this problem. These frames were designed for the spinning of natural fibres. Man-made staple fibres are spun on the frames used for the natural fibres they most closely resemble. In blends of natural and man-made fibres the combined fibres are spun on the frame used for the natural fibres alone. There are four machines for spinning yarn and these are the flyer, ring, mule and cap.

FLYER SPINNING

The flyer spinner is used to spin linen either wet or dry, certain wool fibres, mohair and other long natural fibres, where better control of such fibres is attained. The flyer frame is also employed for inserting a low twist into rovings of other fibres, during the preparation prior to spinning.

In the spinning of linen, a finer yarn is produced if the roving is wetted in a trough of water placed before the spindle. This binds the fibres closer during the spinning process and gives a high twist to the yarn.

The roving passes through the final drafting rollers, enters the spindle and travels down the hollow arms of the flyer and on to a bobbin. This is held on the spindle and as the spindle and flyer rotate, the bobbin is pulled round by the thread. A twist is inserted between the final drafting rollers and the end of the flyer arm. The thread is wound on by the up and down movement of the bobbin.

RING SPINNING

Ring spinning is employed for worsted and cotton fibres after the preparation processes. The roving leaves the final drafting rollers and passes through a pot eye and traveller on to the bobbin. The traveller is a small ring and can slide around the large ring, which is moving up and down. As the bobbin on the spindle is turned it throws the thread around and the traveller is pulled with it. Twist is inserted and the up and down movement of the ring ensures even winding on.

CAP SPINNING

Wool fibres may be spun into worsted yarn on the cap spinner. This method, like the ring spinner, gives a high production of yarn.

SPINNING METHODS

The roving coming through the final drafting rollers is passed through a pot eye and twisted round by the bobbin as it rotates. This is attached to a plate at the base of the tube, which is mechanically driven. The tube, on which the bobbin is placed, rotates around a stationary spindle carrying the cap. A lifting plate pushes the bobbin in and out of the cap and the edge of the cap guides the thread.

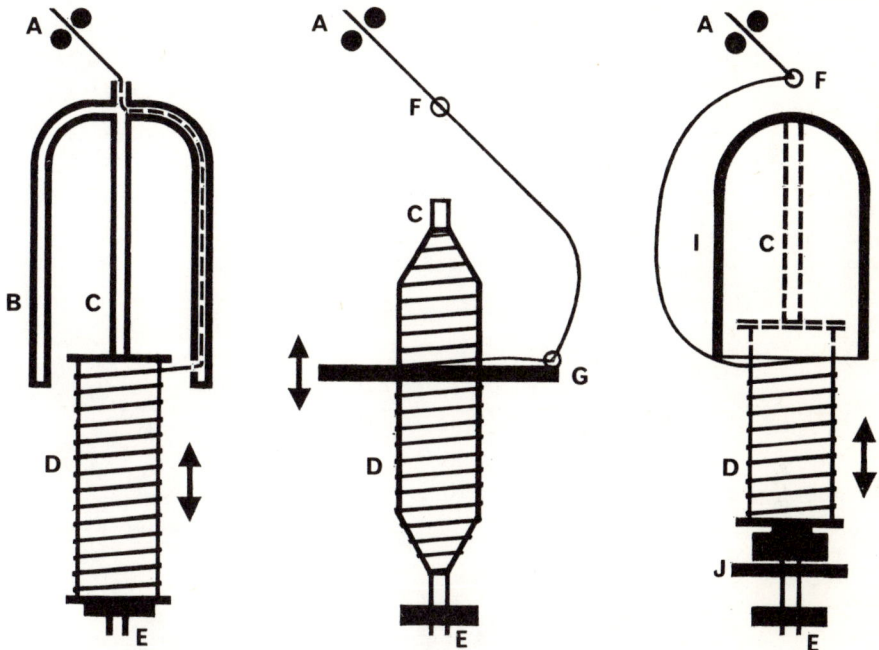

Diagram 7. Flyer, Ring and Cap Spinning. A = final drafting rollers B = flyer C = spindle D = bobbin E = spindle drive F = pot eye G = traveller H = ring I = cap J = plate.

MULE SPINNING

Wool and yarns of a very high count are spun on the mule spinning frame. It is more economical to employ the other methods of spinning for cotton, although a small amount is spun on the mule. The amount of twist inserted in the yarn depends on the speed of the roving from the final drafting rollers to the spindle. A roving released slowly will have the greater twist.

In the woollen mule spinning frame the carriage moves out, the condensed slubbing is delivered by way of the guide (B) and delivery rollers (C). When the carriage is about $\frac{3}{4}$ out, delivery ceases and so the slubbing is drawn out to the desired thickness or count. When the carriage is fully out, the spindle is rapidly revolved and the required twist is distributed along the whole length. The carriage moves in and the spun yarn is guided on to the slowly revolving spindle by faller (F) and tensioned by faller (E). After the carriage reaches the in-end, the whole sequence is repeated.

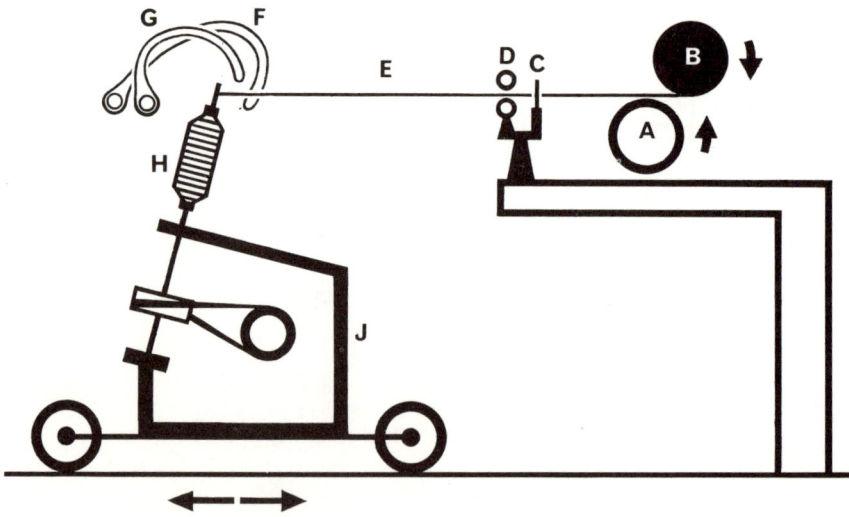

Diagram 8. Woollen Mule Spinning. A = surface drum B = condenser bobbin C = raddle or guide D = delivery rollers E = yarn F = counter faller G = winding faller H = cop on spindle J = carriage.

SPINNING METHODS

FOLDED OR PLIED YARN

Fibres that have been spun by the usual methods of spinning are produced as a single thread or single yarn, and some woollen yarns are used for woven textiles in this form. In order to strengthen or thicken a yarn, two or more single threads may be twisted together; this gives greater strength and evenness than a single yarn of the same thickness. The twisting together of single threads is called a folded or plied yarn. Two single threads twisted together would then be referred to as a 2-fold or 2-ply yarn, and four singles twisted together would make a 4-fold or 4-ply and so on. It is possible for the yarn with four threads to be of the same thickness as the 2-ply yarn, the four singles being very much finer.

With this method of folding threads together a great variety of yarns may be obtained, such as fancy yarns. A folded or plied yarn will withstand the strain imposed on it during the weaving process better than a single yarn, which has a tendency to fray, if there is a weakness in the twisting of the fibres.

TWIST OF YARN

In order to produce a yarn of good evenness throughout, it must be twisted evenly in the doubling frame. The twist of yarn is the number of turns about its axis in a unit length, which is expressed in turns per inch or turns per metre in a known diameter. The twist is controlled mechanically and the higher the twist the finer and harder the yarn will be. Woollen yarn receives very little twist as its frictional properties would not withstand a high twist imposed on it. Cotton can have a very high twist. A yarn which has been soft spun may have as little as 6 to 10 turns or twists per inch and a hard spun as many as 20 or more.

Those yarns to be used for the warp will usually have more twist than weft yarns, to take the strain imposed upon them by tension during the weaving process. The twist in a yarn may be inserted in two ways, clockwise which is comparable with the letter S and anti-clockwise as in the letter Z. By holding a piece of yarn straight from the spool, in a vertical direction, it is possible to discover by unwinding which type of twist it has. The direction of the twist in yarn can be very important, and a length of material may be ruined by a yarn with the wrong twist for the type of weave employed.

Diagram 9. S and Z Twist.

A yarn with a high amount of twist becomes rather harsh and unwieldy and if used for curtain material would not drape well. It would be more suitable for an upholstery fabric, whereas a slightly twisted yarn in upholstery would after some abrasion begin to show loose fibres.

A simple method of finding the number of turns or twists per cm in a yarn can be carried out by inserting two pins into the yarn, 1 cm apart. One pin is held firmly with the head at the top while the other pin, starting with the head also at the top, is twisted over and over. Each time the head of the pin comes to the top it will be counted as one turn or twist. This is continued until the threads of the yarn are separated and are lying parallel to one another.

SPINNING METHODS 43

FANCY OR NOVELTY YARNS

These yarns are produced in many ways, either by blending differently coloured fibres and spinning them as one yarn: by printing or dyeing a pattern on to a sliver or yarn and by adding small spots or neps of coloured fibres and twisting them in with the threads. A great variety of yarns are also obtained by twisting together two or more threads which are different in softness, thickness and colour. The amount and direction of twist may be altered so that the threads twist together unevenly. Some yarns are formed by one thread snarling up, or having lumps, loops, knops, thick and thin places.

Fancy yarns may be made from all natural fibres, all man-made fibres, and a blending of the two groups. Other yarns can have one thread in natural fibres twisted with a thread of man-made fibres. Yarns achieved by combining the two groups in this way are most effective in the woven cloth as the natural yarns tend to be lustreless and the man-made usually have a sheen.

All fancy yarns used in woven fabrics give texture and colour. They are very decorative, but would not produce a hard-wearing cloth if used alone, and are combined with straight hard-wearing yarns. Fancy yarns are more expensive to produce and by using them with straight yarns the cost of fabrics is reduced.

The heavier fancy yarns are used only in the weft, as in a warp they tend to become unravelled with the strain imposed upon them. Fine fancy yarns may be used for the warp as they pass easily through the loom during the weaving operation. All fancy yarns have their own name and this is given because of their appearance, the way in which they have been produced, or from the fibres used in their manufacture.

The fabrics produced from fancy yarns combined with straight yarns are used for all purposes, dress, upholstery, curtaining and vision nets.

SPIRAL YARN

This consists of a 2-ply thread which is twisted with a thick soft twisted thread. The latter is entered into the twisting frame more quickly than the 2-ply thread.

GIMP YARN

A yarn similar to the spiral, but it is harder than the latter and is finer in size. An easily recognisable yarn as it is formed in the shape of the letter S joined together at the top and bottom continuously. It has two fine threads twisting in opposite directions and holding in position a third thread, forming the S loop.

SNARL YARN

Three threads are used to form the yarn. One is entered as a slack thread and hard twisted so that it twists up and produces snarls. The other two, twisting in opposite directions, hold it in position. This yarn has a very rough feel.

KNOP YARN

Two threads are twisted together, with one at regular intervals being given in very rapidly so that it is wound round and round the first thread in the form of a hard knop or lump. The two threads are then twisted in the reverse direction with a third which acts as a binder.

CLOUD YARN

This yarn is produced by two fine threads twisted together, and at intervals a portion of a thick soft roving is introduced and twisted with them. The threads are entered rapidly with the roving so that there is much less twist in the thick parts of the yarn than the thin parts.

TUFFLE YARN

A yarn made from several threads twisted together to form lumps in continuous succession throughout the yarn.

SPINNING METHODS

NEP YARN
This is made from two threads twisted together and small pieces of roving forming flecks or neps are not thoroughly twisted in and tend to lie more on the surface of the yarn. The nep is added in the final stage of carding.

CURL OR LOOP YARN
This yarn consists of a fine foundation thread and a soft spun thick thread, which forms loops at regular intervals by being rapidly given in to the twisting frame. A fine binder thread holds the loops in position.

BOUCLÉ
A term referring to yarns having a curl, loop or knot effect. The yarns are made from two or more threads twisted together.

GRANDELLE YARN
Two or more differently coloured threads are twisted together. Bright colours such as orange, blue, scarlet and lemon may be twisted with black or white.

MARL YARN
A yarn produced by spinning from two differently coloured rovings. The colours are less distinctive than a grandelle yarn. Two marl threads are sometimes twisted together to form one yarn.

NUB YARN
A rather irregular thread containing small spots or nubs produced in a similar manner to knop yarn.

FLAKE YARN
A coloured ground thread, which has large patches or flakes of coloured or white roving twisted in.

CHENILLE YARN
This yarn has a central core of threads and fibres projecting all round. It is mainly used as a weft yarn.

LUREX

A metallic thread produced in a variety of colours by coating thin sheets of aluminium on both sides with a non-tarnish thermoplastic resin. The sheets are then cut into fine filaments.

THE SIZE OF YARNS

The size or thickness of a yarn is referred to as the count, or denier. These are calculated on the relationship between the weight and length and indicate how many units of a standard length make up 1 lb. in weight.

The unit is usually in the form of hanks or skeins of yarn, which is a continuous length of yarn wound on to a reel of a standard diameter, thus forming an O-shaped bundle of loops. The total length of yarn in the hank varies according to the type of fibre. The weight of 1 lb. is chosen as a standard measurement and the number of hanks needed to make up 1 lb. in weight indicates the yarn count. The finer the yarn, the more hanks needed to make 1 lb., therefore a fine yarn will have more yards to the pound than a thick yarn. A fine yarn has a high count number and a thick yarn a low count number. The sizes of cotton and worsted yarns are calculated in this way.

When the yarn is plied this is also expressed in the count, for instance a 2-ply cotton yarn with a count of 12 would be referred to as a 2/12's.

The system of wool sizes is confusing, as each part of the country has its own term. In Yorkshire the size is described in 'skeins' and is calculated on the number of skeins each containing 256 yards required to make up 1 lb. of yarn. The Scottish wools are termed 'cuts'. This system is calculated on the unit of 300 yards of yarn required to make up 24 oz., which is equivalent to the Yorkshire system. In the West of England, the term is 'snaps'. For linen the size is expressed in 'lea', with 300 yards wet spun yarn to the 1 lb.

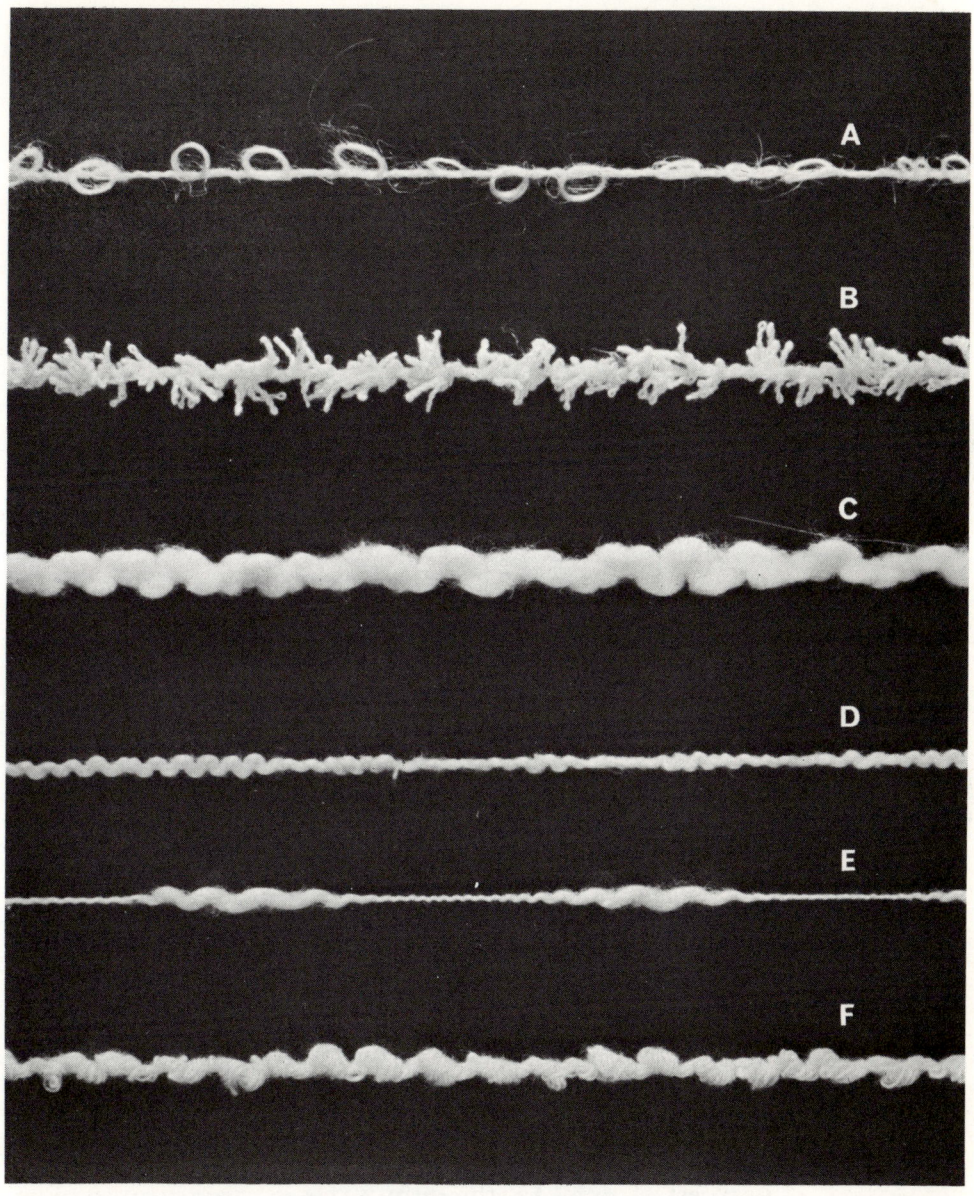

Plate 1. Fancy Yarns (a) Loop Yarn (b) Snarl Yarn (c) Gimp Yarn (d) Gimp Yarn (e) Flake Yarn (f) Tuffle Yarn.

All these figures refer to a singles or 1's yarn making up the weight.

Denier denotes the size of filament yarns such as silk, nylon and rayons. The denier thickness is achieved by calculating it usually as 9,000 metres of yarn and that weight in grammes gives the denier. The thicker the yarn the greater its weight and as the size of the yarn increases so does the denier. An example of this is found in nylon stockings where the 30 denier is coarser than the 15 denier. The denier size may be made up of several filaments together and one example would be a yarn of 120 deniers made up of thirty threads. This is conveyed by 120/30.

Man-made staple fibres are described in the terms of the spinning method used, such as on a machine for cotton. This yarn would then have CC added after its count to show that it was spun as a cotton.

The Tex system is an international system for all fibres, calculated on the weight in grammes of 1000 metres. When the yarn is plied the count is given first (R) followed by the number of strands in the ply. A 2/6's yarn would be translated into the Tex system as R12 Tex/2. The two singles of 6 count would each measure 1000 metres weighing 6 grammes. Plied together they equal 1000 metres weighing 12 grammes and the count may be written as Tex 6 × 2 or R12 Tex/2.

SIMPLE TESTS TO IDENTIFY FIBRES

BURNING TESTS

It is possible to identify fibres by subjecting them to a flame. The fibre is first brought near to the flame and a synthetic type fibre will shrink back on approaching it. This helps in identifying them, and other phenomena are the smell, fumes and residue left after burning. With a piece of cloth it is advisable to pull out a few strands of yarn from the warp and weft and burn them separately. In many fabrics yarns of different fibres have been used, which can

SPINNING METHODS   49

make it a little more difficult to construe the burning test accurately. It is possible to sort them out into different groups for similarity of appearance before burning.

WOOL: Melts before igniting. It burns slowly with an odour of burnt feathers or hair and the flame is easily extinguished. The residue left is gummy and brittle.

COTTON: Burns rapidly with a yellow flame and an odour of burnt paper. The residue is a grey ash.

LINEN: Burns rapidly, but not as fast as cotton, with a yellow flame, and gives off an odour of burnt paper.

NETT SILK: Melts and splutters with an odour of burnt feathers or hair. Small globules are formed and when they are cool they set firm and can be easily crushed.

LOADED SILK: Chars and does not ignite. It turns a reddish brown or dark grey and gives off an odour of burning wood.

ACETATE RAYON: Melts when held near a flame, leaving small black globules; closer to the flame it splutters and burns slowly. An all acetate rayon sample of fabric will show hard edges and give off an odour of burnt sugar when applied to a flame.

VISCOSE RAYON: Burns rapidly and readily with a yellow flame and an odour of burnt paper. It burns similarly to cotton.

NYLON: Does not burn but curls up and melts when a flame is applied to it, leaving small globules which become hard when cool. The odour given off is of celery.

CAUSTIC SODA TESTS

A solution made from caustic soda and water can be used to identify yarns. The fibres or yarns are boiled in this solution. Cotton, linen, jute and hemp turn yellow or brown, silk and all animal fibres dissolve and rayons swell.

CHAPTER 5

The Dyeing of Yarn

Hanks; Cheeses of Yarn; Warp Beam; Cake and Cop

Coloured fabrics can be achieved in various ways, by dyeing the staple fibres and slivers, or by adding dye to the spinning solution of man-made yarns. The yarn may be dyed prior to weaving or the cloth can be dyed after it has been woven.

The dyeing of yarn can be done in various forms, with the yarn in hanks, on cheeses (a cylindrical pack of yarn wound on to a perforated spindle), as cakes (a flat round package of yarn on a perforated spindle), and as a warp wound on to a large perforated cylinder.

HANKS

The dyeing of hanks of yarn may be done by hand or

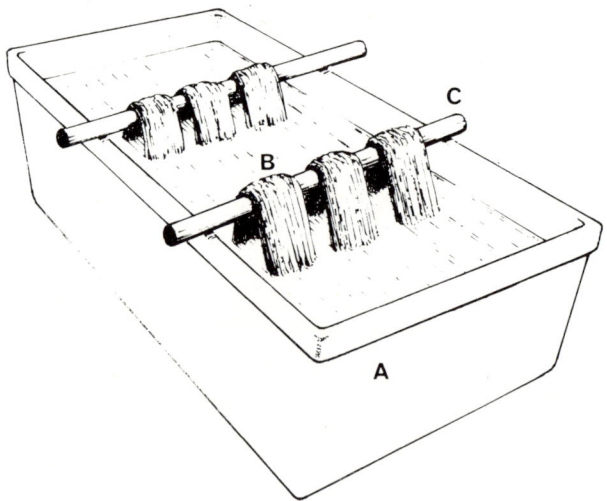

Diagram 10. Hand Dyeing of Hanks. A = dye vat B = hanks C = rods.

THE DYEING OF YARN 51

machine. In hand dyeing large tanks called vats are used in which the hanks are suspended from rods laid on top of the vats. The rods may be of metal, glass, wood or rubber and are often U-shaped to allow the hanks to be handled below the level of dye liquor. The vat is wide enough to allow the hanks to move freely in the liquid; this ensures even dyeing with no danger of patchiness in the yarn. The operator holds the hanks on the rod with one hand and turns them with the other so that the hanks are constantly moving in the dye liquid. This method of dyeing is only used for special colours, exclusive fabrics or tapestries where the depth or matching of a colour may be much more controlled by the dyer.

In the machine dyeing of hanks the output of dyed yarn is much higher than by hand. The rods are replaced by rollers that are driven forwards and backwards. The movement of the rollers permits the hanks to move constantly in the dye liquor so that evenness of dyeing is achieved.

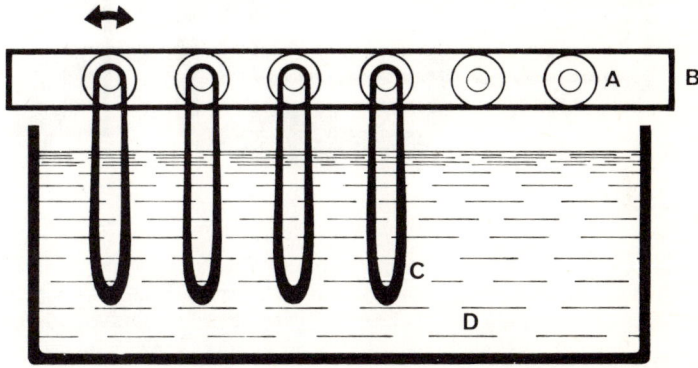

Diagram 11. Machine Dyeing of Hanks. A = rollers B = frame C = hanks D = dye vat.

Another method of machine dyeing hanks is to keep them stationary, stretched between rods with the dye liquor circulating around the vat by a pumping action.

CHEESES

The cheeses, which are short tubular reels of yarn wound on to a spindle, are inserted into a container with a circulation pump. The container resembles a rather large domestic pressure cooker. The cheeses are mounted on special carriers for some containers and in others the cheese spindles act as a circulating system. In another method the cheeses are removed from their tubes and transferred to long spindles mounted to the carrier. A pump drives the dye liquor up the spindles and allows the dye to circulate through the yarn either from inside to outside or vice versa according to the circulating system.

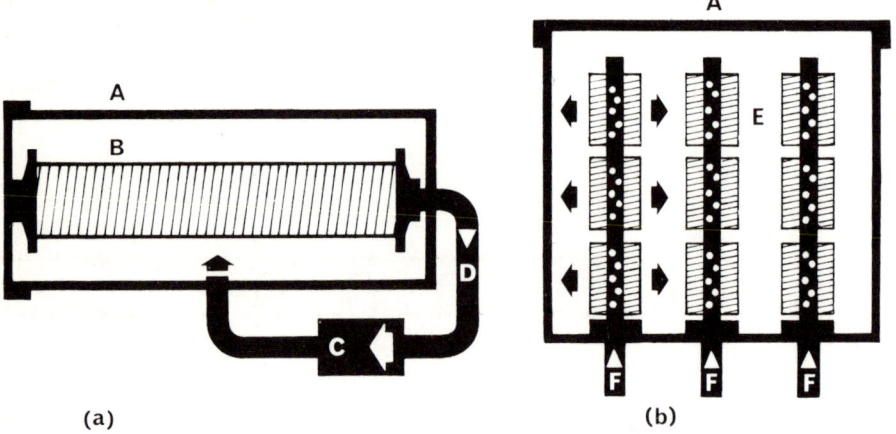

Diagram 12. Beam and Cheese Dyeing, (a) beam dyeing, (b) cheese dyeing. A = dye vat B = warp beam C = pump D = circulation pipe E = cheese of yarn F = perforated spindles.

WARP BEAM

It is very often more economical for a manufacturer to dye a warp (the vertical threads in a length of cloth) on the beam. The yarn is wound on to a large perforated cylinder which is similar to the beam or roller on the back of a loom and inserted into a container connected with a circulation pump. The dye liquor is forced by the pump around this

THE DYEING OF YARN

container and through the beams, giving an evenness of colour throughout. When dyeing is completed the yarn is subjected to various processes to fix the dye while still on the beam. The yarn is dried, rewound on to a warp beam and placed in the loom.

CAKE AND COP

A cake of yarn is similar to a cheese but much rounder and flatter; it is also wound on to a cone instead of a tube. A cop of yarn is also on a cone, but in this instance it is much thinner and taller than the cake. These two are dyed by the same method as cheese dyeing, where the material is stationary and the liquor is moved. With this method of dyeing a good pump and tight fittings on the container are essential for the penetration, depth and evenness of dyeing.

CHAPTER 6

Woven Fabrics

The Sett of Warp and Weft; Warping; Weaving of Fabrics; Drafts and Peg Plans; Plain Weave; Hopsack. Analysis of Cloth for Draft and Peg Plan

Woven fabrics are composed of vertical and horizontal threads which interlace to form a fabric. The longitudinal threads on a fabric constitute the warp and those threads traversing the width of the cloth constitute the weft. The warp threads are known individually as 'ends' and the weft threads individually as 'picks'. These threads forming the cloth may interlace in any manner that the designer requires, taking into consideration the everyday use the fabric will be subjected to. For example it would not be practical for a fabric to have long threads floating or not interlacing when it was planned as an upholstery material.

In order to give a woven fabric a good edge, extra ends are placed in the warp for the first and last few threads. The ends are doubled and may be of a different and stronger yarn to the ones used for the rest of the warp. These are called the selvedges and they provide added strength to the warp for the weaving process, and a firmness to the edges of the cloth for the finishing operations.

THE SETT OF WARP AND WEFT

The sett of warp means the number of ends per inch (or per cm) used for a fabric. This depends on the type of fabric required, and the same yarn spaced close or open can give a quite different cloth. The count and ply of yarn determines the number of ends to be used when a warp is planned. A yarn with a low count and too many ends per inch would produce a hard unwieldy cloth. A high count yarn, and insufficient ends per inch, could produce a weak cloth, which would eventually become misshapen after wear.

The term picks per inch denotes the number of weft threads used to weave 1 inch of cloth. This number depends on the count or denier of the yarn and the weaves that are employed. In a closely woven cloth, the number must be the same for each inch, or horizontal striping will occur.

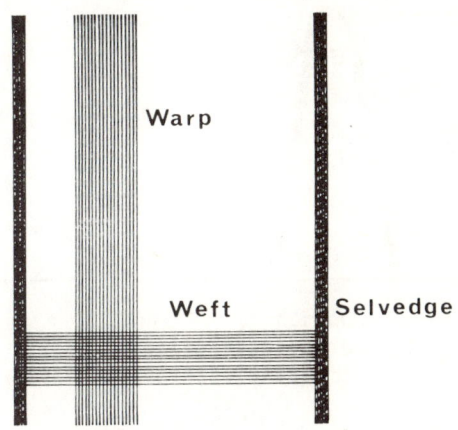

Diagram 13. Direction of Warp and Weft.

WARPING

All the yarns to be used for the warp are on cheeses, cones or bobbins placed on the protruding prongs of a frame called the creel. To avoid variable yarn tension with the rotating of the cheeses, effective tensioning devices are employed. The threads are led through a comb so that they are all parallel to one another and are wound on to a cylindrical roller (warp beam). A warp may be prepared in sections known as portees, which are later combined.

The warp has to be tightly stretched during weaving, and a considerable amount of strain is imposed on the threads. In order to strengthen them, most warps are usually sized by an impregnation of starch or dextrin, mixed with fatty and mineral substances. A recent development is a plastic type adhesive to replace the starch. The surface of the warp yarns must also be as free as possible from surplus fibres, and sizing also overcomes this problem. The size is removed from the cloth after it has been woven.

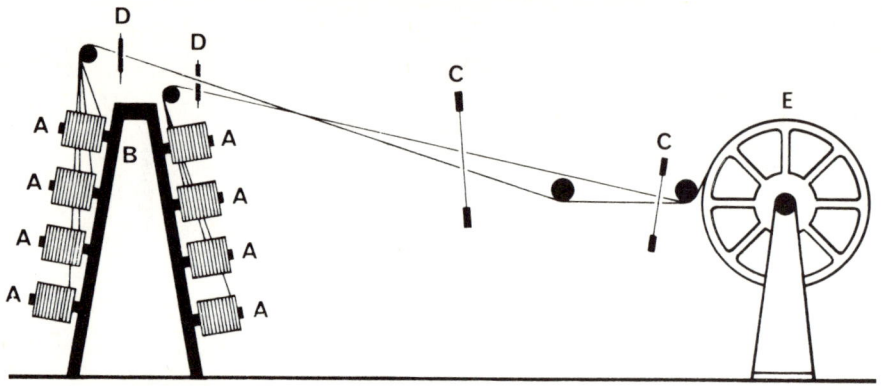

Diagram 14. Warping. A = bobbins of yarn B = creel C = comb D = spindles E = warp beam.

WEAVING OF FABRICS

All woven fabrics are produced on a loom. The warp threads are held under tension and the principal movements of the loom are shedding, picking and beating.

The warp is wound on to the warp beam, situated at the back of the loom, and passes under and over a pair of lease rods. The ends are threaded through centre eyelets (mails) on wires (healds) with one heald for one warp end. These wires are on rectangular frames called shafts or staves. From the healds, the ends are then entered through the reed, which is similar in appearance to a comb, but is enclosed on all sides. The spaces between the teeth are referred to as dents, and it is through these dents that the warp ends are drawn. The completed warp is then attached to the front of the loom and weaving commences.

A number of shafts are raised and the others are either

WOVEN FABRICS

left in the original position or lowered. This parting of the warp threads is referred to as shedding. The weft travels between the two sets of threads from one selvedge across to the other. After the insertion of one pick of weft (picking), the reed on its forward movement pushes it into position (beating) and then moves back. The warp ends are then lowered and another set is raised for the weft to be inserted. The three motions of shedding, picking and beating, closely follow each other and each time the reed beats the weft up against the already woven cloth (fell of cloth).

The weft yarn is wound on to a bobbin or pirn, contained in a shuttle, which is an oblong double casing, resembling a small boat. The shuttles are housed in a shuttle box at the side of the reed and are automatically released one at a time as required, to be knocked by the picking stick through the shed.

The warp automatically moves through the loom from the warp beam, and is woven into cloth, and wound on to the cloth beam at the front of the loom.

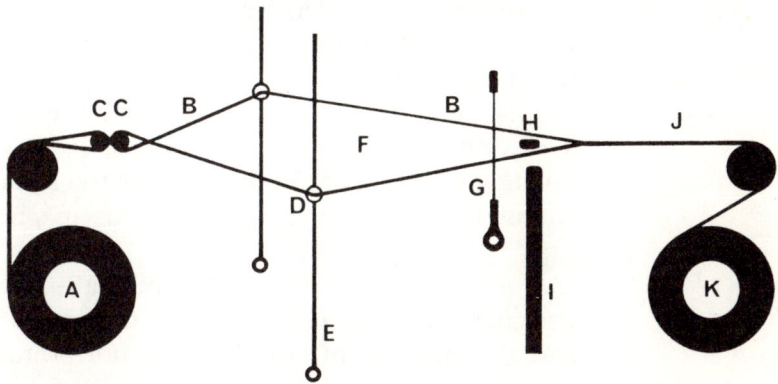

Diagram 15. Loom. A = warp beam B = warp C = lease rods D = mails E = healds F = shed G = reed H = shuttle I = picking stick J = woven cloth K = cloth beam.

DRAFTS AND PEG PLANS

All looms, except the Jacquard loom, have shafts or staves

containing the healds through which the warp threads are drawn. These warp ends must be entered in some definite order, so that they will interlace in the correct sequence with the weft. This interlacing in a recognised order of warp and weft will give the repeat of the design. A design repeat may be only a few ends and picks, repeating many times across the width and length of the cloth, or it may be several inches, depending on the type of loom used and the design required.

Each individual warp end entered on each shaft is shown on a draft. A graph paper is used for this purpose called design or point paper. This is ruled in vertical parallel lines, as shown at A (Diag. 16), with the spaces between the lines representing the ends, and in horizontal lines, with the spaces in between representing the shafts, as shown at B. When these two sets of lines are put together they form squares as shown at C.

The choice of drafts is so wide that the designer must work out the draft and weave together for a successful design. The draft depends on the number of shafts to be used, and all threads which lift alike may be drawn through healds on the same shaft.

The point paper can be divided into small squares of 8×8, with thick lines called 'bars' between each set. The dividing of the paper by bars allows the squares to be easily counted to find the number of ends in one repeat of the draft.

If a draft for a loom with eight shafts is being used, the designer will use point paper with eight vertical squares, and the number of squares in a horizontal direction will depend on the amount of ends entered for one repeat. The reading of drafts is usually from left to right, and from the bottom to the top. The first square in the bottom left hand corner is shaft number 1 with the first end threaded through the heald eyelet. This is shown by filling in the square, indicating a warp end, the second square from the left and the bottom is shaft number 2, end number 2, and that is also filled in to indicate a warp end on that shaft. The third square from the left and the bottom shows end number 3 and shaft number 3. This method is continued until all the warp ends in one repeat have been entered on to all the shafts. The draft shown in Diag. 17A is termed a straight draft and the repeat is on eight ends. When the eighth end has been entered on the eighth shaft the repeat starts again with the ninth warp end on shaft number 1.

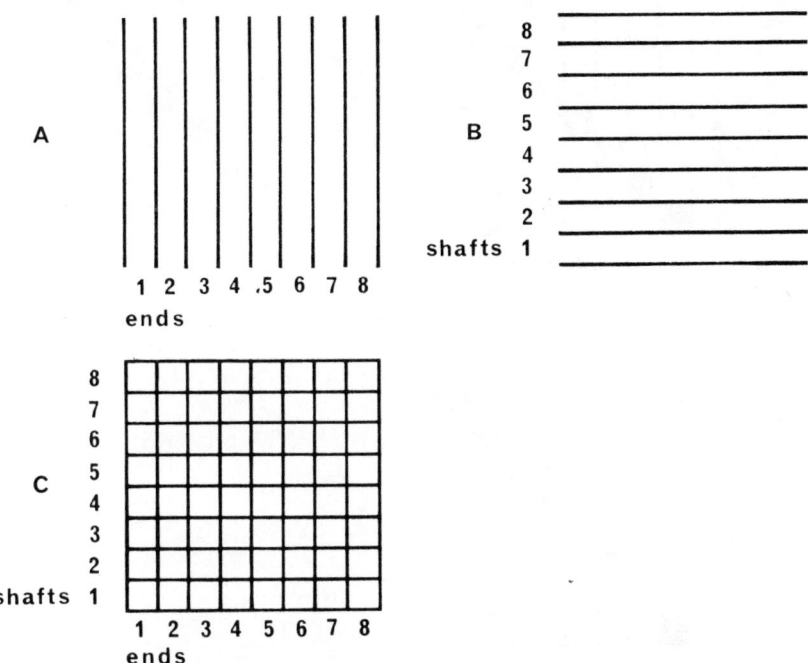

Diagram 16. Draft Point Paper.

If a loom with 16 shafts is used, then two squares of 8 × 8 amounting to 16 small squares counting from bottom to top will be used and the number across from left to right will depend on the repeat.

The most simple drafts are the straight drafts, that is

the ends entered from 1 to 8 consecutively, or 1 to 16 according to the number of shafts used. The centre or point repeat draft (17B), is another simple draft and will give a bigger size of repeat as the design repeats back on itself. If a half circle is woven, the draft repeating back will give a full circle. In a point repeat the ninth warp end on an eight shaft loom would be entered on shaft number 7.

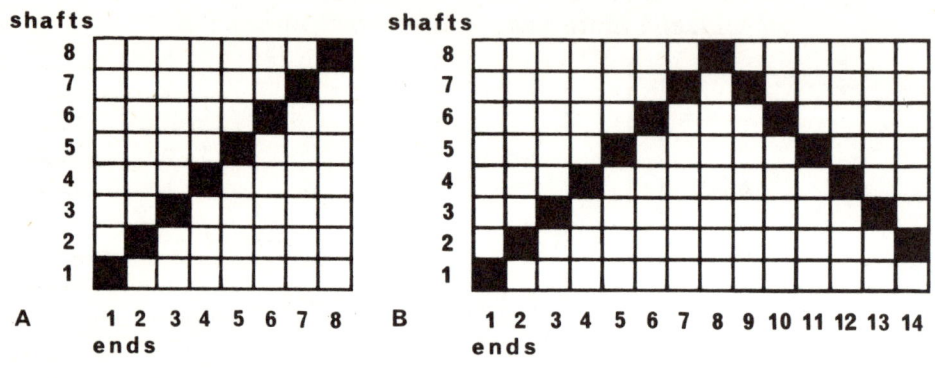

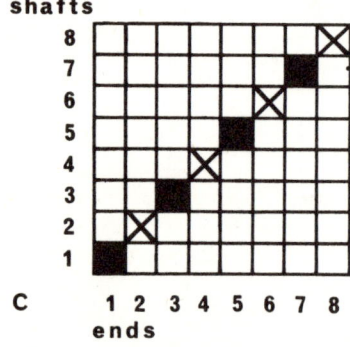

Diagram 17. Drafts. A = straight draft B = point repeat draft C = straight draft for 1 and 1 warp.

With some other drafts care must be taken not to have too many ends entered on one or two shafts, as this will cause breakages in the warp through the lack of freedom to move. Shafts with more ends on them than others are usually arranged at the front of the loom as they are easier to get to should a breakage occur.

For a multi-coloured warp, a striped warp or a warp in two colours, it is possible to show the colour sequence in the draft by means of symbols for each colour. The example shown in Diag. 17C is a warp with one black end and one white end (1 and 1). The point paper shows that the filled-in squares represent the black ends and the squares with an X the white ends. Various symbols may also be used to show the different yarns on each shaft of the loom.

Another method of showing the warp colour order is as follows. All the yarns used in the repeat are written at one side in their colours, count, and ply sizes. The example (Diag. 18) shows that the lemon was used for one warp end, followed by two ends of white, one end of lemon, six ends of white, one black end and so on. This warping order shows clearly and simply the sequence of colours just as a draft shows precisely which end is entered on each shaft.

lemon 2/4's plain cotton	1	1	1
white 2/12's mercerised cotton	2	6	4
black 2/4's plain cotton		1	1
brown 2/12's mercerised cotton			1

Diagram 18. Warping Order.

In order to show the weave, a design or point paper is used to indicate the shaft lifting order or peg plan. For the weave, the vertical parallel spaces between the lines, as shown at 19A, represent warp ends and in the horizontal lines the spaces in between represent weft picks, as shown at B. These two sets of lines put together form squares C.

The two sets of threads, warp and weft, cannot be seen together at the same time, as one thread must be over the other interlacing to form the cloth. The weft thread must go over and under the warp threads and this is shown by filling in a square indicating that the warp end shows and leaving a blank to indicate the weft. By this method it is very easy to show the different weaves, or order of lifting the warp threads.

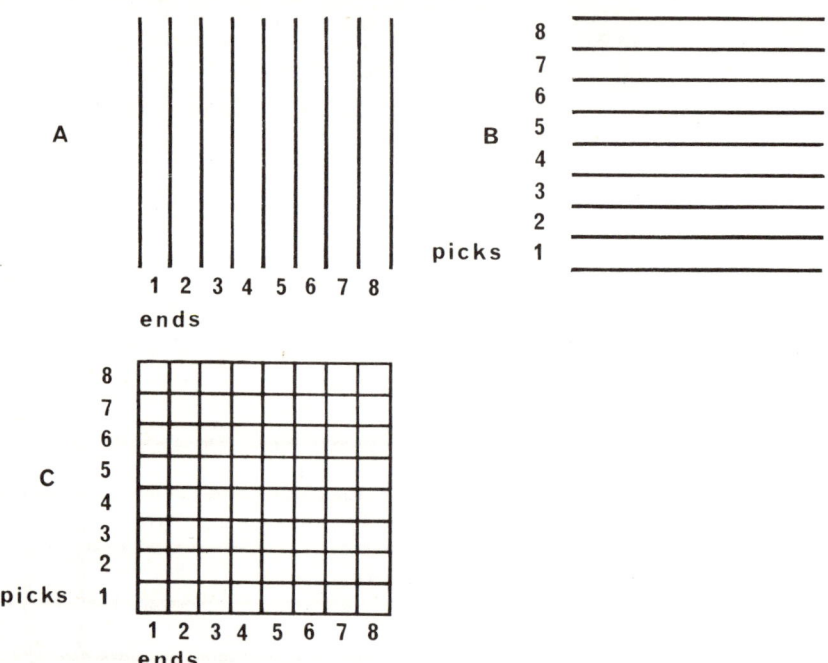

Diagram 19. Peg Plan Point Paper.

PLAIN WEAVE

The simplest weave is the plain or tabby weave, in which the pick of weft travels under the first end of warp, over the second, and under the third. On the next pick it travels over all the ends it went under before. In this weave all the odd numbered warp ends have been raised, forming the shed

WOVEN FABRICS 63

for the first pick. The odd numbered ends are then lowered and the even numbered ends raised to form the second shedding. The point paper of this weave would then appear as a chess board with the black squares as ends and the white as picks (Diag. 20).

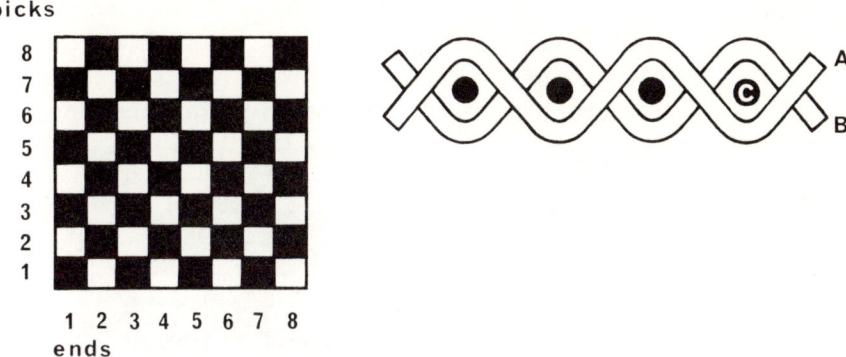

Diagram 20.   Peg Plan for Tabby or Plain Weave. A = 1st weft pick   B = 2nd weft pick   C = warp

The weave shows that on the first pick of weft the ends on shafts 1357 were raised and on the following pick shafts 2468 were raised, each shaft pulling up the warp ends. The number of squares used for the weave depends on the number of shafts and the number of picks for the repeat. It follows that if a draft for eight shafts has been used the peg plan must have eight squares also, to show how each shaft must be lifted. The eight squares are counted from left to right and the number from top to bottom depends on the number of picks used. If a loom with sixteen shafts is used the peg plan requires sixteen squares counting them horizontally, and vertically it will once more depend on the number of picks per repeat.

The shafts which are to be raised are shown clearly on the point paper by filling in the appropriate square. Several different designs may be produced from one draft by using many different peg plans. A straight draft for instance will allow a great variety of weaves but the repeat will tend to be small.

HOPSACK

This weave is similar to plain weave, but the ends and picks are doubled. The diagram shows the double ends and double picks forming a type of matt weave. A variety of fabrics are woven in hopsack weave and these include dress, suitings and furnishings. This strong, hard-wearing weave can also be used as a part of Jacquard ornamentation.

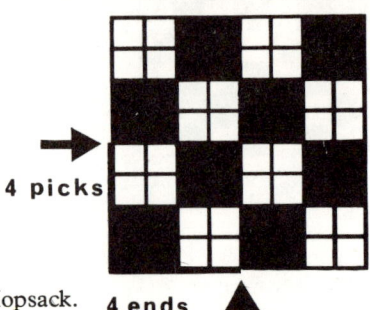

Diagram 21. 2/2 Hopsack.

ANALYSIS OF CLOTH FOR DRAFT AND PEG PLAN

The analysis of a fabric consists of discovering its structure as produced by the yarns, weave, sett, draft and peg plan. The easiest way of analysing a cloth is with a magnifying glass, thread counting or linen glass.

To discover how a fabric was woven and the type of loom used, it is important to find one repeat of the design across the width and length. It can then be marked clearly so that all the area within these marks must be analysed. This can be a laborious and tedious business, as the first step in this direction must be to put every warp end that shows in the cloth on to point paper. Working from left to right across the repeat and from top to bottom or vice versa, examine every pick of weft and fill in a square on point paper where every warp thread shows. Work along each pick of the woven fabric within the repeat, filling in the appropriate squares on the point paper. A needle or pin will be of great use in helping to keep track of the threads.

Once the design of the cloth has been worked out the draft and peg plan can be discovered. It may be difficult to discern from a small sample of cloth which is warp and which weft. If a selvedge is present there is no difficulty, but without one there are certain guiding factors that will help. The warp is usually the stronger yarn and if a fabric is composed of cotton and rayon it would be better to assume that the cotton could be the warp. The number of threads used horizontally and vertically are also a guide and as a rule the greater would generally be the warp. The warp threads are usually of a harder twist and if a single yarn is used with a plied yarn, the latter would also be the warp.

The point paper (Diag. 22A) with the design of cloth is placed below more point paper in order to find the draft. The first vertical line on the design point paper at the left hand side is taken as the first end and on the draft point paper the bottom left hand square is filled in, indicating that the first end is entered on shaft 1 (Diag. 22B) and all the other identical vertical lines will also be on shaft 1. The second vertical line of squares on the design point paper is then examined, and if it is different to the first one, it means it must be on another shaft. This is indicated by filling in square number two on the draft paper (C) and all other identical vertical lines will be entered on this shaft. If the third vertical line differs from the first two, this will be shown as being entered on shaft number three, (D) and the fourth vertical line will be as at (E).

This method is used all across the repeat, taking each vertical line of the point paper and showing the entry on the draft. Once the draft has been completed (F) it is possible to work out the peg plan, which will be as long as the repeat and with the same number of squares as there are shafts. This is placed at the side of the design of cloth.

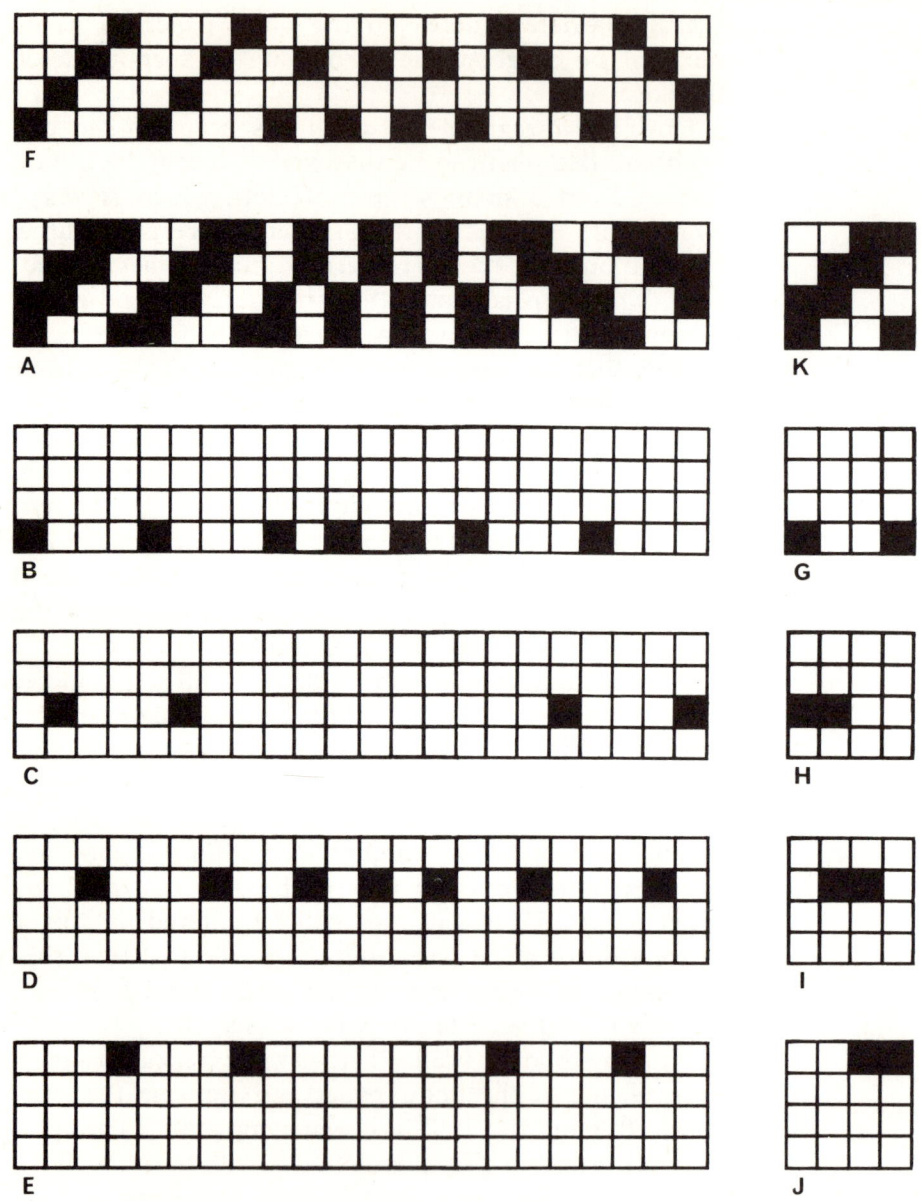

Diagram 22. Analysis of Cloth for Draft and Peg Plan.

The first horizontal pick of the design point paper is examined. All those warp ends that are on the surface have been lifted to allow the weft through. They are traced up to the draft to see which shafts they are on, and then marked in on the peg plan point paper level with the line of the pick (G). The second horizontal pick on the design point paper will show the warp ends on that line and this is also traced up to the draft and marked in on the peg plan (H) and so on for the third (I) and fourth (J) until one repeat of the peg plan has been completed (K).

Cloth analysis of this type can take a long time as the design of cloth must be accurate; any error will make it impossible to work out the draft and peg plan.

NOTE

Since this book was first published, the shuttleless loom has found increasing favour with weavers. A note on machines of this type will be found as an Appendix, page 221.

FROM FIBRES TO FABRICS

CHAPTER 7

Fabrics Produced with Plain Weave

Appliqued Fabric; Azlin; Bag Cloths; Baize; Batiste; Billiard Cloth; Broadcloth; Buckram; Calico; Canton; Cambric; Camlet; Casement Cloth; Cheese Cloth; Chiffon; Crash; Crepe de Chine; Duck; Georgette; Gingham; Glass Cloth; Gloria; Grosgrain; Hessian; Holland; Jaconet; Lawn; Limbric; Mull; Muslin; Nainsook; Ninon; Organdie; Ottoman; Petersham; Pina Cloth; Pongee; Poplin; Pyjama Cloth; Repp; Scrim; Seersucker; Sheeting; Shantung; Taffeta; Tussah; Voile

Many fabrics are woven in plain or tabby weave, as this is the simplest method of interlacing the warp and weft, which can be produced on any loom. The woven fabrics may be very different in texture and strength, depending on the yarn used, the sett of the warp and the picks per inch.

A fabric with a very fine yarn used for the warp and weft can have many ends and picks per inch and give a strong but fine fabric in a plain weave; poplin is a good example of this. The same weave is used for hessian, but the material is totally different, due to the thick, coarse yarn used in both warp and weft. Some fabrics have a closely sett warp in a fine yarn, and by weaving a thick yarn across, the weft is almost completely covered by the warp, which gives a ribbed effect.

Plain weave is the easiest of all textile weaves to recognise, as it can be analysed from only two picks of weft woven through the warp; most other weaves require several picks to show the weave. This may also be used as a background weave for fabrics with spots or decorative stripes; it gives a firm ground to the cloth and enables the spots or stripes to be woven on the surface. They are caught into the background at regular intervals, so that the floating threads

FABRICS PRODUCED WITH PLAIN WEAVE

forming the design will not give a weakness to the fabric. In this method the colours of the spots or stripes may be changed after a number of yards have been woven, thus giving alternate colourways in the most economical manner.

Plain or tabby weave is widely used for all types of woven textiles such as shirtings, dress fabrics, suitings, curtain materials, household textiles and upholstery in both natural and man-made fibres alone, in blends and mixtures.

APPLIQUED FABRIC

A fine woven material which is ornamented by sewing or embroidering an opaque fabric to the surface. The background areas are cut away after sewing, leaving the thin fabric.

AZLIN

A cotton material in plain weave and used for soft furnishings. It is produced in a variety of plain colours.

BAG CLOTHS

Woven in poor quality yarns, giving a rather light and open structure. It is heavily sized to prevent the contents, such as flour, salt, grain, etc. from coming through.

BAIZE

A fabric woven in a woollen yarn. It is heavily felted and piece dyed (see Chap. 18) and the surface has a long pile produced by raising.

BATISTE

A fine, lightweight cloth in plain weave. The majority of cloth is produced in cotton yarn, but a fine linen batiste is also woven. Mainly used for handkerchiefs and underwear.

BILLIARD CLOTH

A woollen cloth made from Merino wool. It is heavily milled (see Chap. 14) with a fibrous finish. The cloth is dyed green or red and used for billiards table or card table tops.

BROADCLOTH

A heavy hard-wearing woollen cloth in yarn made from Merino wool. It is heavily milled and finished. It takes its name from the finished width of 56 inches woven in the loom at 90 inches. It is one of the oldest types of woollen cloths.

BUCKRAM

A coarse cotton fabric in a plain weave. It is piece dyed and stiffened with size. One class of buckram consists of two stiffened fabrics glued together.

CALICO

A term applied to woven cotton cloth in various qualities which are coarser than muslin. Calico is used for various purposes, dress fabrics, blouses, underwear and for printed curtaining.

CANTON

A cloth woven with warp and weft in botany, or with a cotton warp and botany weft. Canton crepe is woven in plain weave with silk in both warp and weft. Canton flannel, a strong twill cloth, is woven in cotton and finished with a raised surface on one side.

CAMBRIC

Woven from fine bleached cotton with a warp of 2/80's at 96 ends per inch and 2/80's to 2/120's cotton at about 80 to 144 picks per inch. It can give a rather stiff, bright finished cloth and is used for summer dresses or with a soft finish for dress linings. Cambrics for embroidery are in a coarser cotton size.

CAMLET

A plain cloth originally woven in camel hair yarns and replaced now by worsted.

CASEMENT CLOTH

A cotton fabric in white or cream with the weft generally

predominating on the surface. It is soft and drapes well when used in curtaining, but is losing its popularity in favour of other fabrics which serve the same purpose.

CHEESE CLOTH (OR BUTTER MUSLIN)

A loosely woven plain fabric used for wrapping cheese and butter as it is very light and soft. This cloth can also be heavily sized and stiffened for use as underlinings. In the soft furnishing trade it is used unsized for quilt inner linings.

CHIFFON

A very soft, filmy material. Made from silk of the finest hard twisted singles and woven in the gum condition, the cloth is degummed after weaving. Chiffon is also made in nylon and rayon. Warp and weft are sett at 100 ends and picks per inch. This material is used for dress fabrics and scarves.

CRASH

A linen fabric in plain weave or in fancy crepe weaves. It has an irregular appearance due to the thick uneven yarns used in the weft. Brown mercerised cotton is sometimes used for the warp in place of linen.

CREPE DE CHINE

A silk material very popular in the 1920's/30's. A plain woven, finely crinkled fabric made from a silk warp and crepe twisted silk weft, woven in the gum condition and afterwards degummed and piece dyed. A rayon imitation is also made. Crepe de chine is used for dresses and blouses.

DUCK

Heavy, strong cotton or linen canvas fabrics used as sail cloth, and for awnings and tents.

GEORGETTE

A filmy fabric with an all-over crepe appearance, woven in silk with very hard twisted threads. The cloth, in plain weave, is degummed after weaving and is piece dyed. Cotton

Georgette is made in imitation of the silk fabric with hard twisted warp and weft yarns. It is also woven in a variety of man-made yarns.

GINGHAM

A cotton fabric usually with a striped warp and weft in the same order so that checks are formed. A fine yarn is used, with many ends and picks per inch to give a firm cloth. Gingham is a popular fabric for clothing and for kitchen curtains as it is hard-wearing, retaining colour and shape after frequent washing and exposure to sunlight.

GLASS CLOTH

Woven in all cotton, all linen or from both yarns in a plain weave. It can be woven in stripes, checks or plain and afterwards printed on. Used for drying and polishing glassware and china.

GLORIA

A strong, firm, cotton fabric used for umbrella coverings.

GROSGRAIN

A plain weave is used to produce a cloth with a prominent warp rib effect. The warp is closely sett and can be in silk, fine cotton or rayon with a weft in silk, worsted, cotton or rayon. Used for dress fabrics and women's lightweight suitings.

HESSIAN

A coarse jute yarn is used for the warp and weft. It is in plain weave with 12 ends and 13 picks per inch. The cloth is fibrous and made in a variety of qualities. It can be dyed but is most often used in its natural colour in the sacking and upholstery trades and as a wall covering.

HOLLAND

A cloth made from linen yarn in 32's lea warp and weft

FABRICS PRODUCED WITH PLAIN WEAVE

with 42 ends and picks per inch. This fabric has its main uses in the upholstery trade as an undercovering.

JACONET

A fine, plain woven cotton cloth which can be soft or heavily starched and calendered (see Chap. 14) to produce a stiff glazed cloth used by undertakers.

LAWN

Woven in plain weave in cotton or man-made yarns. The cloth is very light, fine and smooth and is used for dress fabrics. The warp and weft are sett at 80 to 90 ends and picks per inch.

LIMBRIC

This fabric has an evenly twisted warp yarn, woven with a soft spun lustrous weft which is thicker than the warp. The cloth may have more picks than ends per inch and is used for dress fabrics.

MULL

A very fine soft spun cotton yarn is used for the warp and weft. The fabric is bleached and given a soft finish. It is used for dresses, veils and turbans. Swiss mull has 80's to 100's warp and weft yarns sett at 100 to 120 ends and picks per inch.

MUSLIN

The term muslin may be applied to plain woven fabrics made of silk, worsted or cotton yarns. The latter yarn is the most frequently used and it gives a soft, fine, open cloth. The plain cloth may be ornamented with cords, crammed stripes, spots and other designs in extra warp or weft. (Crammed stripes are made by increasing the number of ends per dent in the reed.) In Swiss muslins, spotted effects are produced by embroidering the cloth after it is woven. Muslin is used for dresses, blouses, etc.

NAINSOOK

A fine and light plain woven cloth which has been bleached. Made in different qualities and used for dresses.

NINON

Woven in silk or man-made yarns. It is fine, light and soft and is used for dresses and blouses.

ORGANDIE

Woven plain in a fine white 80's cotton warp yarn and 100's fine white cotton weft with about 80 ends and picks per inch. This fabric belongs to the muslin class and has a translucent appearance. It is wiry, rather stiff and is mainly used for frilling and similar purposes

OTTOMAN

Originally woven in all silk, all wool, or a blend of silk and wool, sometimes woven with a cotton weft. Rayons may be substituted for the silk. It is a heavy warp rib fabric and in Ottoman cord the fabric has rib lines running the length of the fabric made with a thick warp and fine weft. A dress fabric which is hard-wearing but not used a great deal now.

PETERSHAM

A belting used for the tops of skirts or for stiffening waistbands. A number are woven side by side in plain weave. The warp is finely sett and a thick weft gives a warp rib structure.

PINA CLOTH

Woven with yarns produced from fibres of the leaves of the pineapple tree. It is very stiff, wiry and similar to a polished cotton fabric.

PONGEE

A plain woven fabric originally made from wild silk in the gum condition and degummed after weaving. Imitations are made in fine mercerised cottons or in rayons.

FABRICS PRODUCED WITH PLAIN WEAVE

POPLIN

This fabric was originally made in silk in both warp and weft. Most poplin today is woven in plain weave from combed and gassed Egyptian cotton. A fine cotton is used for the warp with a thicker weft. Cotton poplin is now mostly mercerised and can be given a moiré finish (see Chap. 14). Irish poplin is made with a silk warp and a fine worsted weft. The fabrics are used for dresses, blouses and shirts.

PYJAMA CLOTH

Woven in various qualities. The yarns used may be silk or cotton. In cotton yarns the warp is closer sett than the weft and can have broad or narrow stripes of colour. The fabric can be brushed to bring up the fibres on the surface.

REPP

A warp rib structure in which rib lines are formed across the cloth. There are several types of repp fabric. In one the ribs are developed by using thick and fine threads alternately in the warp and weft and by passing the thick ends over the thick picks while the fine ends are highly tensioned. Another type of repp has a very fine warp closely sett and a thick weft. An alternative method is with a fine warp and thick weft using several weft threads in the same shed to form strong lines here and there. Repp fabrics can be woven in all cotton, all worsted, or with a cotton warp and worsted weft. This fabric is used as curtaining, loose covers, cushion covers and for upholstery in a heavy quality.

CORKSCREW REPP

Plain woven with a fine warp and a thick spiral weft which gives an irregular appearance to the rib lines.

SCRIM

A loosely woven cloth in different qualities made from cotton yarn. The cloth can be stiffened, pasted to another cloth and used by milliners for hat shapes.

SEERSUCKER

A cotton fabric in plain weave woven by using a warp with alternate stripes of loose and tight tension, fed into the loom from different warp beams. This difference in tension produces crinkled stripes in the woven cloth. The main uses of seersucker are in women's and girls' blouses and dress materials. It is also used for tablecloths and as a bed covering.

SHEETING

Woven in fine linen or cotton yarns in plain weave. Coarse sheeting has a condenser weft and in flannelette sheets the surface is raised to bring up the fibres. Bolton sheeting is in 2 and 2 twill weave.

SHANTUNG

A rather rough fabric woven from Tussah silk yarns. The fabric in its natural brown colour contains imperfections such as lumps or slubs due to the rough quality of the silk. A cotton imitation is made and the weft yarn contains slubs at intervals which are soft unspun yarn. Man-made yarns are also used for Shantung dress fabrics.

TAFFETA

A fabric used for dresses or linings and originally woven in silk. Manufactured now in rayons and synthetic yarns in various qualities in plain weave. Shot taffeta or Chameleon taffeta is woven in contrasting colours of warp and weft in equal proportion. The fabric appears to be one colour from one position and another colour from a different view.

TUSSAH

This fabric was originally woven from silk in light brown, fawn or natural colours. A mercerised cotton is woven and dyed to imitate the silk cloth and imitations are also produced with man-made fibres. Tussah fabrics are used for dresses and blouses.

VOILE

A lightweight, cool, plain woven fabric used for dresses and blouses. Woven in a hard twisted cotton for the warp and cotton weft which has been combed and gassed to give a smooth cloth. Voile may also be ornamented by crammed stripes and extra warp and weft designs. A percentage of man-made fibres blended with the cotton in some fabrics assists in the quick drying of the material.

CHAPTER 8

Fabrics in Twill, Satin and Sateen Weaves

Barathea; Blankets; Botany Twill Cloths; Box Cloth; Coutil; Denim; Drills; Dungaree; Estamene; Felt (Woven); Flannel; Flannelette; Frieze; Foulard; Gabardine; Galatea; Habutai; Imperial Cloth; Kersey; Linsey; Maud; Melton Cloth; Serge; Tartan; Ticking; Tweed; Twills; Herringbone and Chevron; Satin; Sateen. Fabrics in Satin and Sateen Weaves: Amazon; Atlas; Beaver Cloth; Dimity; Doeskin; Duchesse Satin; Habit Cloth; Japanese Satin; Luvisca; Madras Shirting; Satinet; Silesia; Venetian

Fabrics with plain weave have been dealt with in a previous chapter and it has been shown that warp and weft on point paper appear as a chess board. In twill weaves the interlacing of the warp and weft causes diagonal lines to be formed in the cloth. This will be dealt with in more detail later in this chapter. The point paper shows how the diagonal line is formed in the cloth by this method of interlacing. The weave shown in Diag. 23 has a ratio of two warp to two weft, moving out one and up one square or end each time: this is called a 2 and 2 twill.

Diagram 23. 2 and 2 Twill.

FABRICS IN TWILL, SATIN AND SATEEN WEAVES

BARATHEA
A fabric usually woven in a twilled hopsack or broken rib weave in silk, worsted or man-made yarns and used for clothing.

BLANKETS
Thick and heavily milled fabrics woven in woollen yarns, or wool and cotton, and in man-made yarns alone or blended with wool. The sizes of blankets vary and include 80 × 100 in. (203 × 254 cm) and 84 × 95 in. (213 × 140 cm). Blankets are woven in plain weave or 2 and 2 twill and the surface is brushed or raised to give a fibrous surface.

BOTANY TWILL CLOTHS
Woven from botany quality worsted yarns in various weights. The weaves are 2 and 2 or 4 and 4 twills. The fabric is smooth and used for men's and women's suitings.

BOX CLOTH
A thick heavily milled woollen cloth in 2 and 2 twill for coats and suits.

COUTIL
A strong cotton cloth woven in 3 end warp face twill to form a narrow herringbone stripe (see twills). It is piece dyed and usually made up into corsetry.

DENIM
A strong, hard-wearing fabric usually made in 3 and 1 twill weave with the warp in brown, blue, or red woven with a white weft. The colours are fast to washing and the fabric is produced for overalls, jeans, skirts and summer suiting.

DRILLS
Warp faced cotton fabrics woven in different twill weaves. They are used for tropical suitings.

DUNGAREE
> Very similar to denim and is woven in cotton yarns, but the warp and weft are in dyed yarn. A hard-wearing fabric for overalls.

ESTAMENE
> A worsted cloth which has been woven in 2 and 2 twill or 3 and 3 twill. It is piece dyed, milled and finished with a rough fibrous surface. A fabric for the clothing trade.

FELT (WOVEN)
> A fibrous faced woollen cloth, which has been felted to such an extent that the fibres are matted together to obscure the weave.

FLANNEL
> Woven in plain or twill weave, this cloth is soft and warm. English wools are used in the ordinary qualities and merino for the better quality of cloth. Flannel is milled and raised for some qualities and in others the surface is left with a limited amount of fibrous face. The cloth is used for trousers, blazers and suits, usually in grey. White tennis flannels are frequently woven in worsted and finished with a clear surface to the cloth. Molleton flannel is dyed in delicate colours.

FLANNELETTE
> A cotton cloth in a plain or 2 and 2 twill weave. The surface is finished with a fibrous surface in imitation of wool flannel and this is produced almost entirely from the weft.

FRIEZE
> A woollen fabric woven in 2 and 2 twill with a rough fibrous surface. The cloth is heavily felted and in some cloths the surface fibres brought up are laid in one direction. In other fabrics the nap (see Chap. 14) is rubbed into small curls.

FABRICS IN TWILL, SATIN AND SATEEN WEAVES

FOULARD
A silk fabric in 2 and 2 twill, used for dresses and blouses.

GABARDINE
A fabric for raincoats, suits and coats. The weave employed is mostly 2 and 2 twill, but some qualities are woven in 2 and 1 twill. The yarns are cotton, wool, and cotton with wool. Gabardine suiting fabrics are much softer and have worsted for the warp and weft. Terylene gabardine is now being produced for raincoats.

GALATEA
A fabric woven in 2 and 1 twill or 3 and 1 twill in simple stripe patterns for nurses' uniforms.

HABUTAI
A general term for many Japanese silk fabrics which are woven in plain, twill or fancy weaves. The cloth is soft and fine woven in the gum condition and the gum removed by boiling. It is piece dyed in various colours.

IMPERIAL CLOTH
A coating fabric woven in 2 and 2 twill. It has fine worsted yarn for the warp and weft and is piece dyed. Imperial serge is very similar and used for women's suitings. It is looser woven and softer than the Imperial cloth.

KERSEY
A heavily milled coat material, woven from Cheviot and cross bred wool in 2 and 2 twill and finished with a fibrous surface.

LINSEY
A coarse fabric with a cotton warp and a blended yarn of cotton and waste wool for the weft. The weave is plain, or 2 and 2 twill. It is woven in stripes across the width and used for aprons.

MAUD
: The term for checked woollen travelling rugs woven in different tones of grey.

MELTON CLOTH
: A woollen, heavily milled cloth in which the fibres are raised on the surface. These fibres are then cut, reducing them to a short, dense pile. The weave is usually plain or broken 2 and 2 twill and the cloth used for coating.

SERGE
: A twill cloth, hard-wearing and made from worsted or woollen yarn. Used for suitings, school uniforms and coats.

TARTAN
: A Scottish fabric woven in 2 and 2 twill in wool or worsted. The cloth is elaborately coloured in check designs and used for kilts, shawls or plaids. The different clans have their own particular tartans. The width of the cloth is 27 or 28 inches and the length is pleated to form the kilts, so that the pleats are across the width of the material.

  Imitation Scottish tartans are manufactured in wide fabrics and can be woven in all wool, wool and cotton or blends of wool with man-made fibres. These fabrics are made up into coats, suits, shifts, skirts and trousers.

TICKING
: A strong rather stiff fabric woven in twill in white with a black warp stripe. The cloth may be all linen, all cotton or a combination of both. It is used for pillow covers and was formerly in use as a mattress covering but has been replaced by other fabrics now. It is hard-wearing and makes an excellent curtaining and wall covering.

TWEED
: The term tweed describes a robust woollen fabric in simple weaves such as plain, 2 and 2 twill, herringbone and chevron (see twills). It is made from a fairly coarse fleece with a

medium length staple. Tweed is hard-wearing, pliable, elastic, with a firm weave and slightly fibrous surface. The tweed that is probably the best known is Harris, which is one of the most durable. Cheviot yarn woven as tweed cloth is usually light and springy. The weaves used for tweed material have a good balance of warp and weft and this gives a hard wearing material. Donegal tweed is a rough cloth in plain or 2 and 2 twill and is given very little milling as a finishing process. Bannockburn tweed is a Scottish Cheviot cloth woven with a thread of solid colour alternating with a thread composed of two differently coloured threads twisted together. The warp and weft are in the same order of yarns. Tweeds are used for jackets, suiting and coats.

In upholstery, an all tweed fabric has a smoother finish than the clothing tweed, and is only used for easily upholstered furniture. A warp of cotton or linen with a tweed weft reduces the cost of the material and keeps it in better shape for covering purposes. The weaves are simple, as complicated ones would not show effectively. Lurex or rayon are sometimes added for extra texture or decoration.

TWILLS

The twill order of interlacing produces diagonal lines in the cloth. These weaves give a cloth greater weight, substance and firmness than can be produced by similar yarns in plain weave. Twilled effects are made in various ways and in a simple twill the weft moves one thread outward and one thread upward on succeeding picks. A twill, unlike plain weave, cannot be made with only two threads in the warp and two in the weft, but on any number exceeding two. A simple twill weave is complete on the same number of ends as picks.

The twill lines are formed on both sides of the cloth in a diagonal line from the right or the left, from one selvedge to the other selvedge. The steepness of the diagonal line depends on the type of twill and the kind of weft yarn used. The direction of the line on one side of the cloth is the opposite to that on the other side. If a warp float (long warp thread) predominates on one side of the cloth, a weft

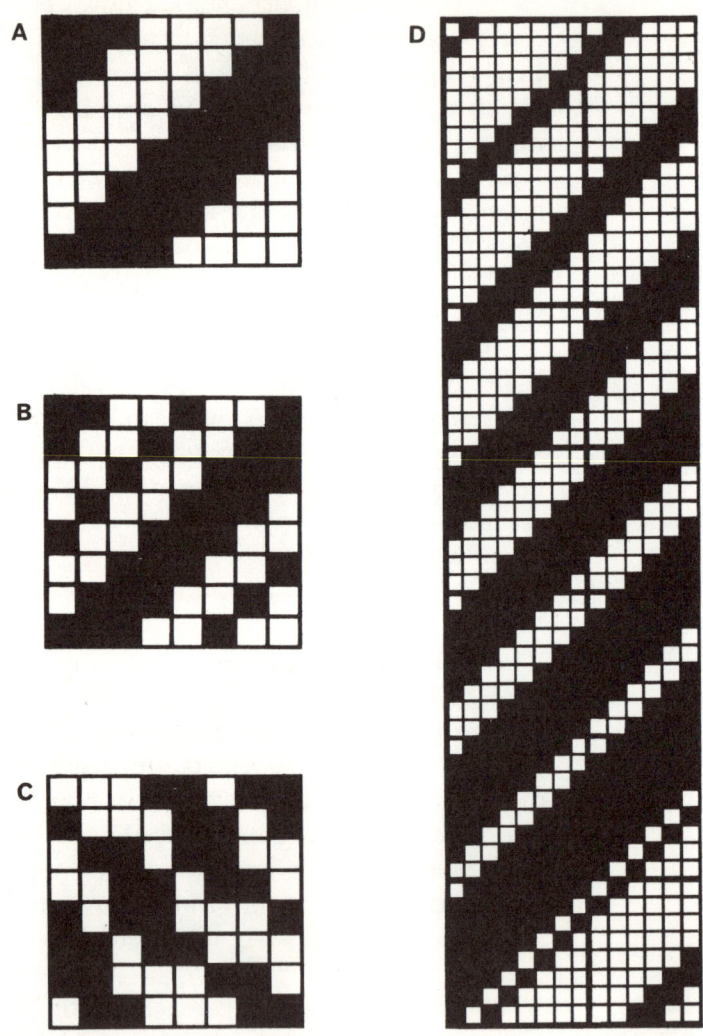

Diagram 24. Twill Weaves. A = regular twill B = regular twill C = broken twill D = graded twill.

FABRICS IN TWILL, SATIN AND SATEEN WEAVES

float will predominate on the other side. In some twill patterns the diagonal line can be changed by grading from a weave showing mostly warp through to a weave with the maximum of weft showing.

There are many types of twill weaves called regular, fancy, zig-zag and broken which give interesting fabrics. These weaves have great use in all kinds of fabrics such as clothing, upholstery and soft furnishings, either as the complete peg plan or as a part of a design combined with other weaves (see Jacquard, page 104).

HERRINGBONE AND CHEVRON

In herringbone and chevron weaves, the twill line is alternated from left to right and right to left, so that a zig-zag line is formed throughout the cloth. The diagonal line can be long or short before reversing, and can be lengthways or widthways on the fabric. The difference between the two weaves is the point of intersection for the twill line. With herringbone, the lines do not completely join together as they do in the chevron weave. This can be seen in Diag. 25.

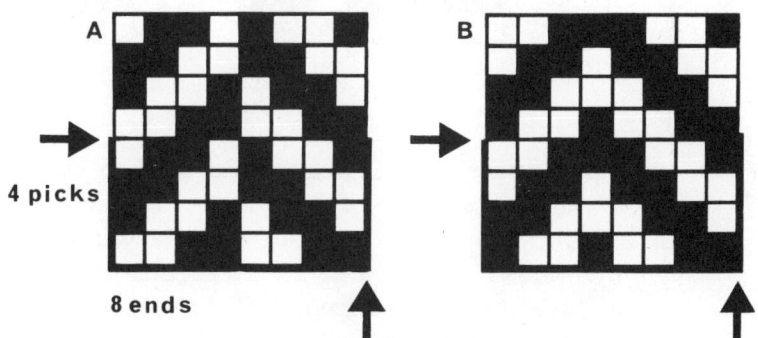

Diagram 25. Herringbone and Chevron Weaves. A = herringbone B = chevron.

SATIN

The satin weave is an extension of the twill weave and the interlacing of warp and weft gives a definite face and back to the fabric. This is achieved by a spacing out, so that the twill lines are not so obvious. The least number of ends that form a satin weave is five and this is called a five end satin.

For satin cloth, the warp is much finer and more closely sett than the weft, and double-faced satins are made with one side a different colour to the other. The back of satin is always sateen, as the warp predominates on the satin side and the weft on the sateen side. Satin weave can be referred to as warp sateen.

Satin cloth is used for dresses, coats, curtains, cushion covers and in good qualities for upholstery. These cloths can be woven in cotton yarns, rayons, synthetic yarns alone or in mixtures.

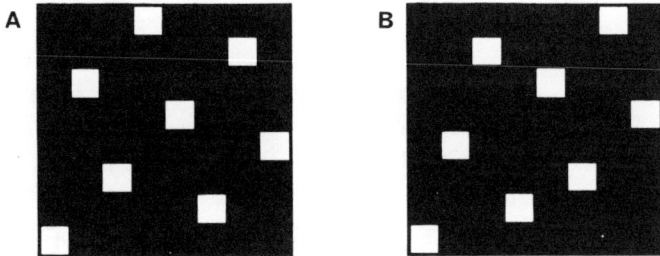

Diagram 26. Satin. A = regular satin B = irregular satin.

SATEEN

In pure sateen weaves the surface of the cloth consists almost entirely of weft floats. In the repeat of the weave the weft passes over all but one warp end. There are two kinds of sateen weaves, the regular and irregular. The latter are entirely free from twill lines, giving them an advantage over regular sateens.

The sateen weave is employed in Jacquard woven fabrics and sateen cloth is manufactured in many different qualities. Sateen cloth can be sold undyed and bleached, or dyed in various colours, for use as curtain linings or as curtaining after being printed on, and as dress materials.

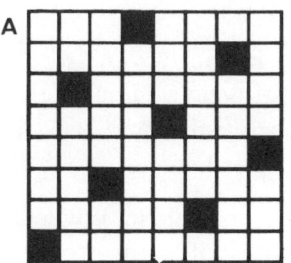 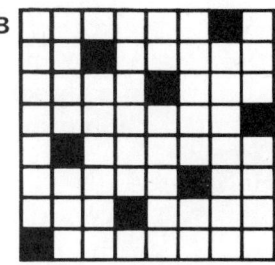

Diagram 27.   Sateen. A = regular sateen B = irregular sateen.

FABRICS IN SATIN AND SATEEN WEAVES

AMAZON

A fine dress fabric with a worsted warp and woollen weft. It is usually woven in a 5 end warp satin with more ends than picks per inch. The cloth is lightly milled and raised to give a fibrous surface but not enough to conceal the fine twill effect of the weave.

ATLAS

A rich fabric made from silk or man-made yarns and woven in an 8 end warp satin weave. It is used for dress fabrics and as linings when the weft is of cotton.

BEAVER CLOTH

A coating material which has been heavily milled and raised. It is woven in a variety of weights in single or double satin weaves.

DIMITY

This type of cloth, woven in 4 thread warp and weft sateen weaves, is used for bed covers. The material is usually formed in stripes equal in size so that the pattern is the same on both sides. The usual size of stripe is four threads each of warp and weft face weave with broader stripes forming a border.

DOESKIN

A very fine woollen fabric woven from high quality merino in a 5 end warp satin weave. The cloth is milled and raised and given a dress face finish and used for dresses.

DUCHESSE SATIN

A very rich and lustrous fabric woven in silk or man-made yarns in various types of satin weave. The warp is very fine and is usually sett at 360 ends per inch, with a thicker weft at 92 picks per inch. This satin is used for dresses.

HABIT CLOTH

A woollen suiting material woven in a 5 end warp satin. It has a fine warp and a thicker weft and the cloth is given a dress face finish after weaving.

JAPANESE SATIN

A fine satin cloth woven in silk in a 5 end satin weave for dresses and blouses.

LUVISCA

A cloth woven in twill or broken sateen in a coloured or striped cotton warp and rayon weft.

MADRAS SHIRTING

A fine lightweight cotton material used for shirts, blouses and dresses. The foundation cloth is plain and it is ornamented with crammed silk or rayon stripes in sateen weaves.

SATINET

Fabrics that are woven to imitate silk satin in mercerised cotton or rayons are called satinet. They are mainly used for dress fabrics.

SILESIA

A cotton lining fabric woven in twill or sateen. It is dyed, heavily starched and finished with a glossy surface and sometimes printed with stripes.

VENETIAN

A cotton lining cloth woven in an 8 end warp satin weave, with more ends than picks per inch. The warp is of good quality cotton, either two-ply or singles yarn. The cloth is piece dyed, mercerised and given a lustrous finish.

CHAPTER 9

Fabrics Woven on the Dobby Loom

Bedford Cord; Distorted Weft; Double Cloth; Folkweave; Honeycomb; Huckaback; Spot Designs; Piqué

The dobby loom allows simple weaves and limited designs to be produced mechanically in a short space of time. The dobby can carry any number of shafts up to 32, but the number mostly employed is 4 to 16. Each individual warp end is entered through the healds according to the draft necessary for the design. The selection of the shafts is controlled by the dobby mechanism and raises them in a prearranged sequence according to the peg plan.

There are various types of dobbies all performing the same function, but with slight variations in the mechanism and the method of pegging the design. In some, the peg plan is embodied in rows of lags, which are thin strips of wood with a series of holes bored through the surface. There must be as many holes as there are shafts, one hole for one shaft. Pegs are placed in the holes according to the peg plan and indicate a heald up or down, depending on whether the loom has a rising or falling shed. Each lag, with its row of pegs, is one pick of the weave and a great number of lags may be used, joined together by the first and last to form an endless chain.

On one type of dobby machine, each peg comes into contact with a bar and pushes it forward. This engages a top hook with a knife blade, and as the knife rises, it takes the hook up. At the same time a lower hook is raised. The shaft is attached to the lower end of the hooks and as the hooks move up they automatically raise those warp ends. The completion of this movement is to return the hooks to

DOBBY HANDLOOM WITCH MECHANISM

Diagram 28. Dobby Loom. A = lag B = peg C = pattern barrel
D = bar E = top hook F = lower hook G = top knife blade
H = lower knife blade I = heald J = warp end.

the original position, with the bottom hook resting on a stationary blade and the top hook disengaged from the top blade. If the shaft is not required for the next pick the hooks remain in this position.

The lags and pegs are placed on a pattern barrel that revolves, bringing into contact the pegs for each pick of weft. The mechanism is placed above the loom and the pattern barrel to one side. This type of loom is largely being replaced by other looms, which allow a quicker production of cloth, and have more in common with the Jacquard.

In this type of dobby, the lags and pegs have been replaced by a tube of thin card, which has the weave cut on it in a series of holes. The mechanism with the card is placed at the side of the loom and is operated by a number of needles and hooks. As the card revolves on a cylinder, it is in close contact with the needles. The holes in the card allow the needles to drop down and this movement causes the hooks to be disengaged. A knife, travelling outwards, pulls out the hooks which in turn raise the shafts. On a 16 shaft loom there are two hooks for each shaft and these travel outwards alternately.

The movement of the loom is much quicker than with the other type of dobby. The distance between each row of holes on the card is very small, and it takes less time for each hole to come into contact with the needles, than for each lag to come into position.

This type of dobby, although closer in style to the Jacquard loom, cannot weave the complex designs produced on the Jacquard, because of the limitations imposed by the shaft system.

BEDFORD CORD

This fabric is used in a very fine form in the clothing trades for dresses, skirts, trousers, and children's wear. It is usually in one colour, with the cords or stripes lengthways

FABRICS WOVEN ON THE DOBBY LOOM

in the material, the ribs being made prominent by well-defined hollows between the stripes. The weave is usually plain or twill, depending on the purpose for which the cloth is intended. They are always warp faced fabrics and the cloth may be given increased weight, thickness or strength, by the addition of extra ends, acting as wadding or padding in the ribs. In most clothing material the cords

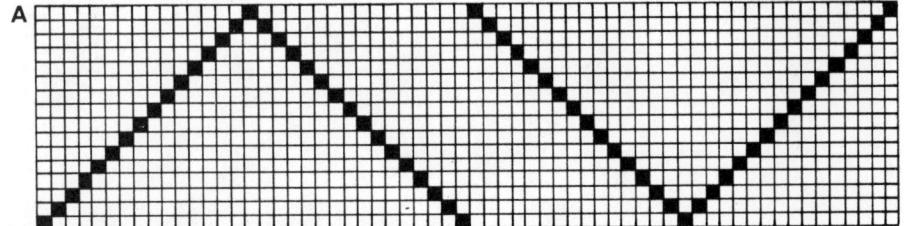

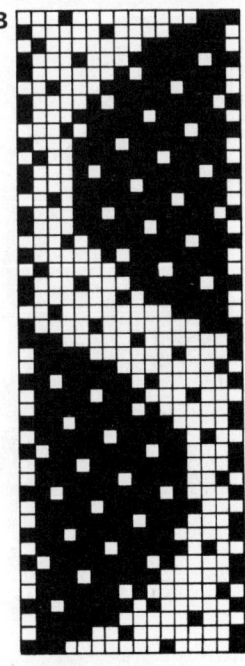

Diagram 29.   Design for the Dobby Loom. A = draft B = peg plan.

Plate 2. Fabric woven on the dobby loom from draft and peg plan at diagram 29.

FABRICS WOVEN ON THE DOBBY LOOM

are very narrow, but they can be made of various widths. The cords can be made in stripes of colour, but this makes the weaving more complex.

The cord is produced by the interlacing of the warp and weft in such a manner that some ends (stitching ends) are woven tight. The other warp threads required to form the cord are woven fairly slack, by interlacing them with only half the number of picks. It is possible to see this on the back of the cloth, where the floats travel across the ribs. The cord may be of any width up to about 1 inch, but the furrows are always two ends only and the unit of repeat is two cords and two furrows.

The warp is sett fairly closely over the ribs, and the two ends, forming the furrows, are not put together in the same dent in the reed. It assists in giving a more pronounced cord if they are split up.

Diagram 30. Bedford Cord. A = weft B = warp.

DISTORTED WEFT

This type of weave is a method of using plain weave and giving it an added interest by a distortion, formed by certain picks of weft. Many fabrics are produced with this type of weave, and it can be used in very fine yarns for dress fabrics and roller towels, or as a larger weave in furnishing fabrics.

The ends are entered to form stripes in the warp. The first stripe is woven in plain weave while the weft threads are allowed to crowd together in the same shed on the second stripe. When the latter is woven in turn, the weft remains in the same shed on the former. By this change over in weave from one stripe to the other, the weft is pulled in such a manner that it distorts and forms a wavy line across the fabric.

DOUBLE CLOTH

This cloth is used mainly for coats and suits, where it is possible to weave two coloured fabrics of the same weight, interlaced so that a reversible material is produced.

The simplest type of double cloth is composed of two series of warp threads and two series of weft threads, with one series forming a face or upper fabric and the other a backing or under fabric. The two series of ends are drawn through the healds in such a manner that one series may work independently of the other.

Diagram 31. Double Cloth. A = upper fabric B = lower fabric.

The face ends are in a definite order with the backing ends and the face picks with the back picks, and separate weaves are required for the two fabrics. If the face picks are woven only with the face ends, according to the weave chosen for that fabric, and the backing picks only with the backing ends according to that weave, two distinct fabrics are formed one above the other. A different weave is produced on both sides and the cloths are joined at the selvedges.

Another method of double cloth is produced by combining

two weaves in one peg plan so that the cloth is interlocked, but it can be one colour on the face side, and another colour on the under or backing side. Some double cloths are woven with the same colour on both sides, if a heavy strong fabric is required.

Double cloth can be woven with one side textured by using fancy yarns, such as mohair loop, and the other side of the fabric in plain straight yarns. Other types may have a check or stripe on the face and are plain on the back. All these materials may be used for coats and suits with one side for the body of the garment and the other side for trimmings. The yarns used for double cloths are wool and worsted alone, or blended with man-made fibres.

FOLKWEAVE

This term refers to coarse, loosely woven cotton fabrics with a woven pattern. They have a peasant and traditional appearance, and can be very colourful. In some fabrics the warp is of cotton and the weft in blends of cotton and fibro, or it can be an all cotton fabric. Folkweave is suitable for curtains, fitted chair covers and cushion covers. It is also woven on the Jacquard loom.

HONEYCOMB

In honeycomb weaves the threads form ridges and hollows, giving a cell-like appearance to the cloths. The warp and weft threads float on both sides, and the rough structure gives an absorbent cloth when the weave is used for towelling. Honeycomb weave is also used for bedcovers, quilts, blankets, dress and coat materials, and a variety of yarns are used, depending on the purpose for which the fabrics are intended.

There are two classes of honeycomb weaves and in one they give a similar effect on both sides of the cloth, and in the other, referred to as Brighton honeycombs, they produce the cell-like formation on one side of the cloth only. The ordinary honeycomb is woven with a point repeat draft and a peg plan in which the warp and weft tend to float. The ridges occur where these long floats of warp and weft are formed and the hollows where the threads interweave in plain weave.

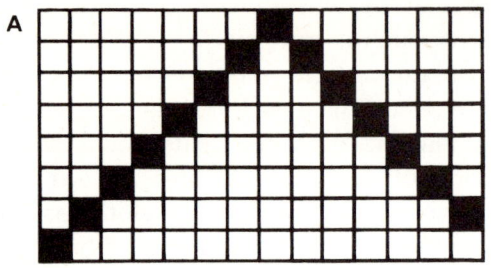 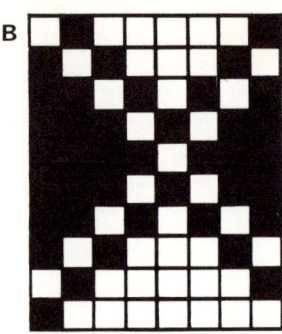

Diagram 32.   Honeycomb. A = draft B = peg plan.

The Brighton honeycomb weaves are quite different in construction and are woven on straight drafts, and the number of threads in a repeat must be a multiple of four. The long centre floats of warp and weft form horizontal and vertical ridges, but two sizes of hollows are formed. This weave produces two large and two small cells as the repeat.

HUCKABACK

These weaves have a plain weave foundation, which gives the cloths firmness and good wearing qualities, and are largely used for linen and cotton towels, glass cloths and some upholstery. Huckaback weaves used for towelling give the cloths good moisture absorbency, due to the comparatively long floats that are formed by the yarns. Huckaback weave can also be used for blouse and dress fabrics in a fine yarn size but the floats are shorter in these fabrics for practical purposes.

FABRICS WOVEN ON THE DOBBY LOOM 99

SPOT DESIGNS

Fabrics woven with a small spot design have been produced in many forms over the last few years. Dress fabrics have been used with small coloured spots woven on to the surface of the material, or fabrics with two colours in the warp and weft with the spots made from one or both of these colours.

A popular spot is produced on a 1 and 1 warp, with one end one colour, and the next end another colour, repeated throughout the warp. The weft may also be in the same two colours or in similar tones and the spots formed by floats from either ends or picks. Another type of spot fabric will weave with a 1 and 1 warp in a fine double cloth, where the upper fabric has a background in one colour and spots in another, and the reverse side is in the opposite colours. For instance a cloth with a black background and white spots will appear as a white cloth with black spots when it is turned over.

PIQUÉ

This type of fabric has an embossed or Italian quilted appearance. The cloth is formed by a series of raised or hollow places, giving whatever pattern or design is required. The face of the fabric is in plain weave and the stitching threads are used for the design or pattern.

The ends interlacing with the weft for plain weave are on one beam on the loom and the stitching ends on another, the former lightly weighted and the latter heavily weighted. The stitching ends, brought up to the face of the fabric, will pull down the cloth at that point, forming a hollow or dent. These stitching ends must be sufficiently strong or bad weaving and faulty cloth will result. By varying the lifting of the stitching ends any pattern may be made and when all the stitching ends are lifted at one time, the fabric is then called a welt.

Another method of weaving piqué is by the introduction

of padding or wadding threads, and this passes between all the ends forming the face of the cloth, and the stitching ends which are left down.

Various classes of piqué are produced, usually in cotton yarns, and very fine piqué cloths are used for women's and girls' dress fabrics, summer suits and coats. In a well woven, good quality of fabric it washes well and keeps its shape. Piqué woven in thicker yarns is also used for upholstery and makes a hard wearing and firm cloth.

CHAPTER 10

Fabrics Woven on the Jacquard Loom

The Cards; The Loom; The Fabrics; Damask; Brocade; Lampas; Brocatelle; Tapestry; Matelasse

The design for the fabric is first carried out in colour on drawing paper, with the width of the loom repeat and the width of the woven cloth taken into consideration. The design is then re-drawn on to Jacquard point paper, which is a strong cartridge paper usually white or blue in colour, and ruled in black squares. The size of point paper varies: 8×10, 8×7, 8×6, 8×5, 8×4 and 8×3 are some of the sizes. A heavy black line is ruled vertically and horizontally after every division of 8×8 etc.

The type of loom and the number of ends and picks per inch, have to be taken into consideration in choosing the size of point paper. The former figure denotes the number of ends, and the latter figure the number of picks. Care is taken in selecting the correct size of paper, as the wrong choice may destroy the proportion of the design. An 8×8 paper would give a cloth with exactly the same number of picks as ends per inch, and 8×5 would give 5 picks lying in the same space as 8 ends.

The number of squares across the top of the point paper is counted equal to the number of ends in a repeat. The repeat depends on the type of loom used, and in a 400's Jacquard the machine controls 400 ends independently for one repeat. If a design is to be woven on a 400's Jacquard loom, the point paper will have 408 squares across the top. The length of the repeat is then calculated and the design is drawn up to that size on the point paper. Some other sizes of Jacquard looms are 300's, 600's and 1200's.

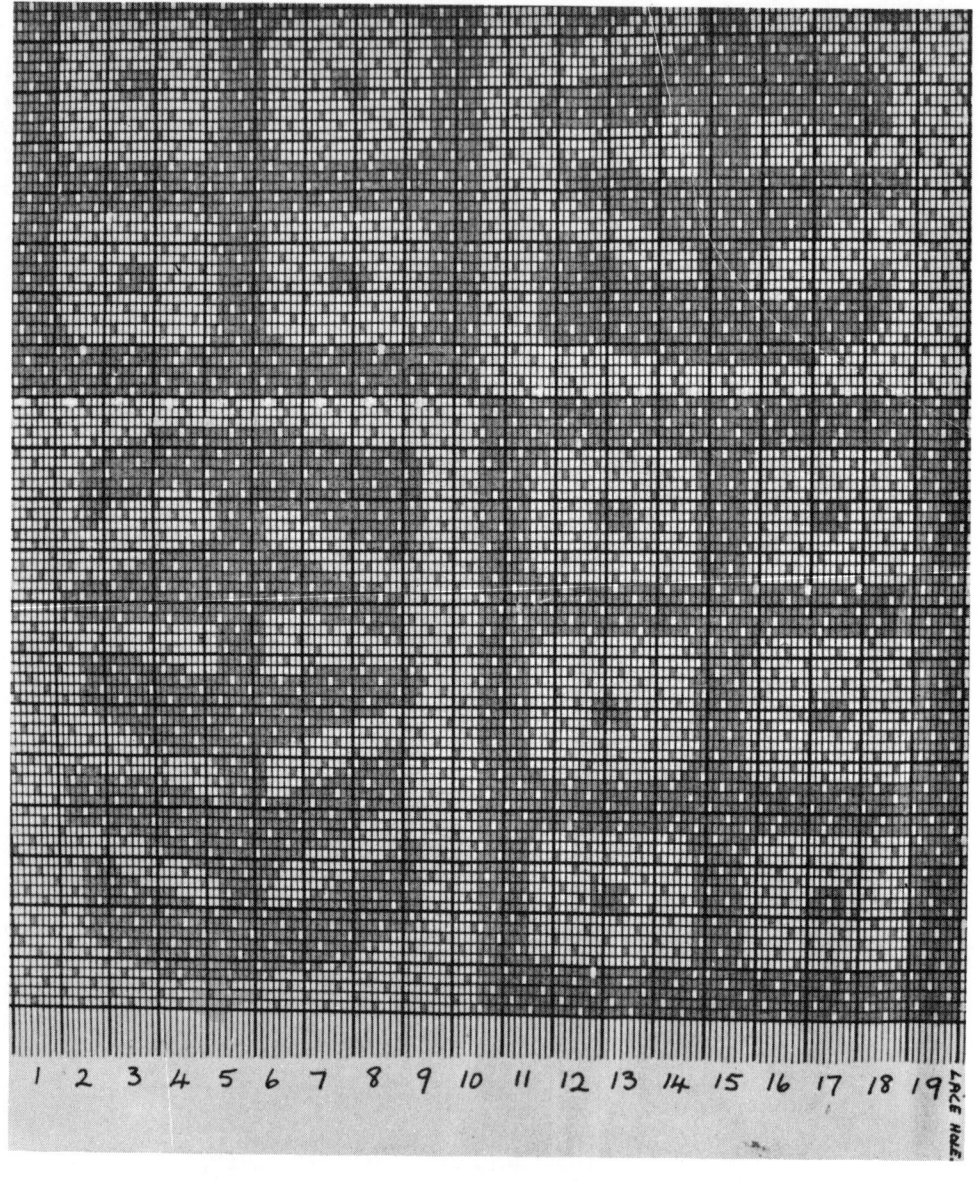

Plate 3. Section of a design on Jacquard point paper in satin and sateen weaves.

Plate 4. Furnishing fabric woven on the Jacquard loom from the design in plate 3. Warp—2/12's black cotton. Weft—gimp yarn.

There are two methods from this point of preparing the design for the loom and the first one is the more traditional method.

After the design has been drawn, the weaves are filled in, and this is usually done in a vermilion colour paint with red squares for warp ends up and white or blue for weft up. In complicated designs other colours are painted in to show the change of weaves.

THE CARDS

Each Jacquard machine has its own size of card, which fits on to it. The cards are oblong in shape, made of cardboard, and are cut according to the point paper design. There must be as many cards as there are picks in one repeat of the design; each pick line is one card. The cutting of the cards is carried out on a machine which punches a row of holes across the width of the card. A hole is punched in the card, where a square of point paper is marked in red. The holes are in rows of eight, to correspond with the blocks of eight squares, denoting ends on the point paper. The first row would be the first 8 ends on the point paper, reading from left to right. The second row corresponds with the second block of 8 ends, or ends 9 to 16. After the first row of holes has been cut, the card is moved on to the next cutting position. Peg holes to fit the card on to the loom are cut at the beginning and end, lace holes are cut at both ends and in the middle.

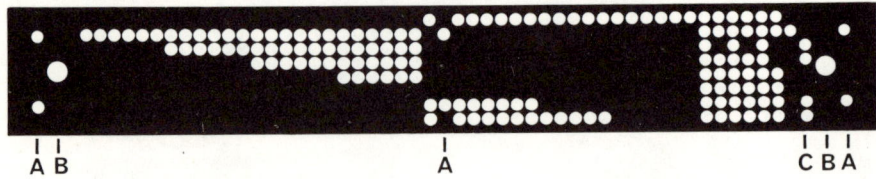

Diagram 33. Jacquard Card. A = lace holes B = peg holes C = selvedge.

The cards, with a series of holes and blanks, are then laced together through the lace holes with a soft tubular tape and form a long strip. Wires are inserted after every sixteenth card and protrude for about 3 to 5 cm. These are used to support the cards when they are placed in a creel or cradle above the loom. After the cards have been placed on the creel, the two ends are passed over the cylinder and tied together to form an endless belt. There may be several thousand cards in one elaborate design, or very few according to the type of design or cloth to be woven.

THE LOOM

There are many different types of Jacquard looms, but they all work on the same principle. A Jacquard machine may be divided into three main sections, the engine, the harness and the mechanism connecting the engine to the loom. The engine is responsible for selecting the warp threads, which will interlace with the weft and form the design as the cloth is woven.

The simplest loom is the single lift, which takes a card with eight holes across the width. Eight horizontal needles on the loom each control a vertical hook, and each needle projects through a needle board facing the pattern cylinder. The cylinder is a four-sided block of wood, the same length and width as a card. It has a number of holes bored into it in width and length, to correspond with all the holes in the cards. A griffe, carrying the lifting knives, is moved up and down for each pick by a driving rod, and when the griffe is in its lowest position, all the hooks will be over their respective knives. At this point, the cylinder carrying the card is pressed against the needles. The pattern cards, with the holes and blanks, come into contact with the needles, which are held in position by a small spring in a spring box. Where there is a hole, the needle penetrates to the cylinder leaving the corresponding hook over its knife, a blank presses in a needle and the corresponding hook is pressed

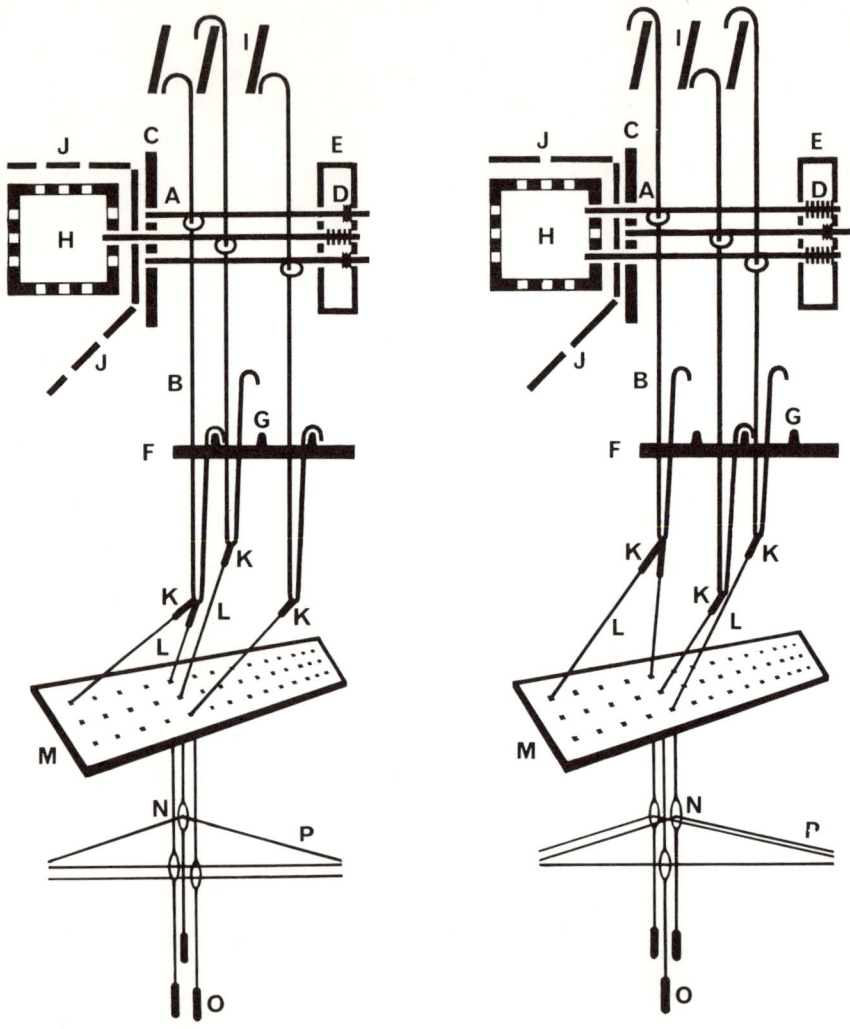

Diagram 34. Jacquard Loom. A = horizontal needles B = vertical hook C = needle board D = spring E = spring box F = grate G = spindle H = cylinder I = lifting knives J = cards K = tail cords L = harness cords M = comber board N = mails O = lingoes P = warp ends.

away from its knife. As the griffe moves up, the hooks over the knives are taken up and those pressed clear are left down.

Connected to the lower end of each hook are the harness cords, similar in appearance to violin strings, and these contain the mails, through which the warp threads are drawn. The harness cords are evenly spaced by being passed through holes in the comber board, with one hole for one cord. The first cord in every repeat of the design is attached to the same hook, the second cord in each repeat to another hook, and so on for all the cords. At the lower end of each harness cord is a small weight called a lingoe. This pulls the cord down after the warp end has been raised and the weft inserted.

After the insertion of each pick, the knives are dropped back, and the cylinder moves away from the needle board and makes a quarter turn, bringing the next card into position facing the needle board. The needles enter the card if there is a hole, so raising the next series of warp ends.

The second method of Jacquard card cutting is a simplification of the first one and allows a more speedy production of material.

The design, having been drawn out on point paper, does not have the weaves filled in, only the shapes. The background weave is left in blank point paper and red paint is used to indicate the second main weave, with other weaves painted in other colours. The weaves are then written down, naming each weave for every part of the design. This method saves a great deal of time, as it is a long and laborious process to paint every weave on the point paper.

Instead of one card for one pick, this has been replaced by one long strip of thick paper or fine card. The weaves are cut on the card, not in a horizontal line reading from the point paper, but by cutting all one weave at the same time in a vertical direction. The card is then re-entered in the cutting machine, to cut all the second weave and the other weaves one at a time.

Sekers Fabrics Ltd.   Plate 5. Section of a design on Jacquard point paper for the fabric in plate 6.

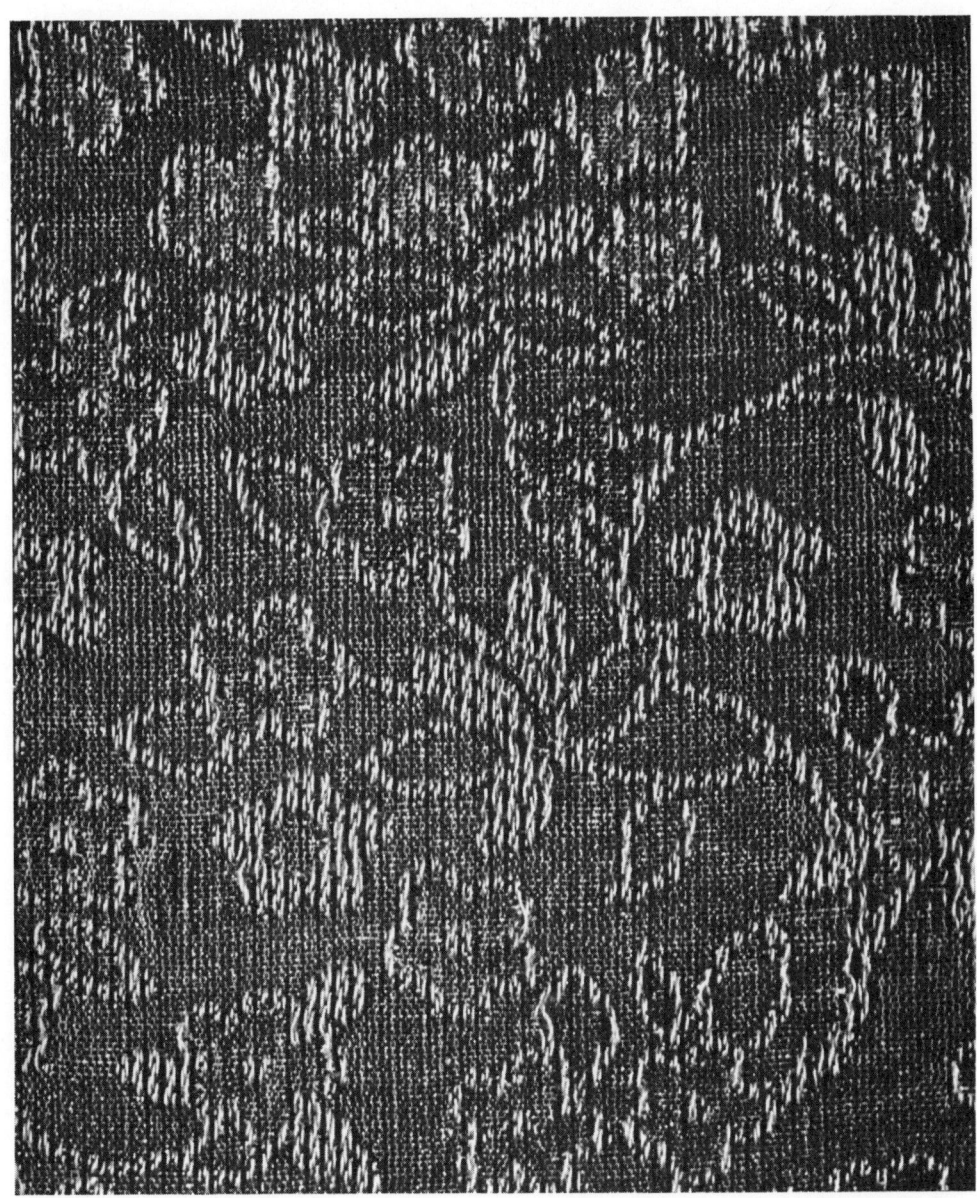

Sekers Fabrics Ltd.

Plate 6. Dress Fabric. Warp—1 end gold Lurex, 3 ends yellow nylon. Weft—orange, yellow and fuchsia rayon. Weave—sateen.

The holes in the card are much finer than the holes used for the separate card machine and are closer together. This means that one continuous card does not take up so much space as the single card system. To hold the card in the cradle or creel above the loom, two small clips are attached at each side, at intervals of approximately 45 cm, and the card is folded across above the clips.

The loom works in the same manner, but the engine has been refined. The needles and hooks are very much finer and the cylinder is replaced by a half circular metal plate. This allows the next pick of the card to come into line quicker than the time it takes in the other method, for the cylinder to make a quarter turn.

THE FABRICS

The Jacquard loom allows complex cloths to be woven, as each end in a repeat can be raised individually. Jacquard woven fabrics may be very ornate in design as a great deal of detail can be achieved. Many designs originally intended for printed fabrics can be successfully converted into Jacquard designs.

The loom needs very little attention. One operator controls several looms and replenishes the bobbins in the shuttles when they are required. The shuttles are contained in boxes at the side of the reed and are placed in the divisions of the box in the correct order of entry through the shed. If the cloth has several colours, the shuttles with each individual colour are automatically released in turn and that part of the design is woven in its correct colour.

The size of repeat, in both dress and furnishing fabrics, depends on the type of loom used and the sett of the warp. For instance, a dress fabric sett at 200 ends per inch on a 600's machine, will give a width repeat of 3 inches. A heavy furnishing fabric, sett at 50 ends per inch, will give a width repeat of 12 inches on a 600's machine. The length of all repeats depends on the type of design and may be only a few inches or as much as 30 inches.

DAMASK

This fabric, woven on the Jacquard loom, is a combination weave, with the background in warp satin and the figure (pattern or design) in weft sateen. Other details may be in twills or plain weave. It was originally a silk fabric, made in Damascus, from which it takes its name.

The cloth is now made in a variety of yarns, cotton, man-made and combinations. The terms single and double damasks are sometimes applied in order to distinguish linen fabrics made in 5 and 8 thread sateens. The best linen damasks have about 50 per cent more picks than ends per inch and would be in a 40's linen lea warp, 70's linen lea weft, with 81 ends and 135 picks per inch.

Damask cloth looks bumpy, but has a flat surface and theoretically is reversible. In some damasks, the effect of pattern is emphasised by the use of warp threads in one colour and the weft in another. Most damasks are self-toned with warp and weft in the same colour, and the design is made prominent by the changes of weave, or the use of one fibre for the warp and another fibre for the weft.

Damasks are used for curtains, table cloths and table napkins. They are not suitable in cotton and rayon yarns for loose covers or upholstery unless closely woven, as the rayon yarns lie on the surface of the cloth and are liable to fray with constant abrasion.

BROCADE

This fabric was originally a heavy, rich, silk fabric with elaborate designs formed by extra threads, or with embroidery. In a true brocade, the design is produced by additional coloured threads in the weft. These float on the back of the cloth and are brought up to the surface when required. The background is usually in satin weave, with the figure in other weaves, twills, etc.

Plate 7. Damask. Width of material 48″/50″. Repeat 10½″ length, 6¼″ width. Woven in 69% cotton, 31% silk.

Many brocades woven today are produced by weaving the extra coloured threads into the back of the fabric. These are only brought to the surface when required by the pattern. The backs of these brocades show stripes of extra weft colours caught into the fabric in certain places to present a smooth flat surface.

Brocades are made in various yarns, all rayon, in combinations of man-made yarns, or with a cotton warp and rayon weft. Various types of brocades are used for curtains, dress and upholstery fabrics.

LAMPAS

A drapery fabric, similar to brocade, and originally an East Indian printed silk. It is now woven with a repp ground and satin-like figure, formed by the warp yarns, and a contrasting figure in the weft yarns. Lampas is principally used for furnishing fabrics.

BROCATELLE

A popular fabric for upholstery as it is firm and hard wearing. The cloth has an additional weft thread of cotton or rayon, that is used to pad the figure, and in a brocatelle of good quality the padding is done with linen. In cheaper qualities, jute is used for the padding and its presence is discernible because of its stiffness and wiriness.

The figures in brocatelle are always in satin weave, combined with other weaves, and the background in twill. The designs follow closely the general types of damask, but in brocatelle the pattern stands out in a raised effect.

Brocatelle is seldon used for curtains as it does not drape well, but it can be used in a fine, thin quality. The yarns for brocatelle are plain cotton, mercerised cotton and rayons; silk may also be used, but the material is rather expensive.

TAPESTRY

A woven fabric that looks embroidered, and was first produced on a loom in imitation of needlework tapestries.

Sanderson & Sons Ltd. Plate 8. Brocatelle. Width of material 51″/52″. Repeat 21½″ length, 25½″ width. Woven in 42% linen, 31% silk, 25% rayon, 2% nylon.

FABRICS WOVEN ON THE JACQUARD LOOM

There are various methods of weaving tapestry cloth on the Jacquard loom, and the effects are obtained by the different weaves on the point paper.

One class of tapestry cloth is made, which has a warp rib surface, and the pattern is produced by employing two, three, or more differently coloured series of figuring ends. Two colours are used for the figuring wefts, one very dark and the other very light. By using three warp colours, combined with two weft colours, eight effects can be produced in any part of the cloth. Additional colours may also be achieved by the planting, or putting in, of extra warp colours. A binding warp and binding weft is also used in addition to the figuring warp and weft.

Other styles of tapestry cloths are produced by figured double and treble plain cloths. These complex structures depend on the application of weaves and the intermingling of colours. The weaves, on all types of tapestry cloths, are arranged so that the most expensive and better quality yarns are brought to the surface. Tapestry cloth is used for upholstery and in a finer quality for curtaining. The yarns employed are cotton and worsted yarns, or cotton alone.

MATELASSE

Matelasse is a double or compound cloth with an embossed or raised appearance. In one type of matelasse the cloth consists of warp and weft, with an extra weft in a coarse count used as a padding. This extra weft padding is usually of cotton condensor. An extra stitching warp holds the padding in position.

Another type of matelasse produced now has a closely sett warp, with one weft yarn, which also acts as the padding. The embossed or raised effect is achieved by the weaves. In all parts of the fabric a double cloth is formed. The back of the cloth is a fine, loosely woven web of warp and weft interlacing together. The weft, which is not weaving with either the surface warp ends or the backing warp ends, is

Sanderson & Sons Ltd.

Plate 9. Tapestry. Width of material 50″/51″. Repeat 16″ length, 12½″ width. Woven in 100% cotton.

Sekers Fabrics Ltd.

Plate 10. Matelasse. Width of material 49″/50″. Repeat 6″ length, 6″ width. Warp—218 ends per inch 150 d. viscose, 15 turns per inch. Weft—76 picks per inch, 30's lea irregular spun viscose slub.

allowed to float between the two cloths and pads the figure. The fabric has a stitching weave outlining the design and this keeps the padding in position.

Matelasse is a hard-wearing cloth and is mainly employed in upholstery, but can be used for curtaining if it is fine enough to drape well. It can also be used for quilts and in a fine quality for evening coats.

CHAPTER 11

Pile Fabrics

Velvet; Genoa Velvet; Terry Velvet; Utrecht Velvet; Figured Velvet; Velour; Velveteen; Figured Velveteen; Corduroy; Needlecord; Designs on Plain Warp Pile Fabrics; Plush; Moquette; Towelling

Pile fabrics is a term referring to all cloths that have threads projecting either in loop form, or as straight fibres from the ground of a fabric. This type of fabric structure gives more thickness, ornamentation and wearing properties to a cloth. The fabrics woven as pile cloths owe much of their effect to the large amount of threads used and the treatment given to the woven cloth in the finishing processes.

Fabrics with a pile can be in various qualities and are used in many different ways, for upholstery, and in the clothing trade for men's, women's and children's wear and household articles, such as towels, curtain materials, cushion covers, etc. The pile is formed either with the warp or weft and can be very simple in construction or very complex.

VELVET

Two warps are used to produce velvet, one for the ground, and one for the pile; these are separately tensioned, but combine to construct the cloth. The ground warp produces with the weft a foundation structure from which the pile threads project.

Two, three or more ground picks are woven in either twill or plain weave, and a shed is formed by raising only the warp threads that have to form the pile. A wire with a cutting blade is inserted and beaten up by the reed in the same manner as a pick of weft. The ground picks are then woven and another wire inserted and beaten up. This process is repeated throughout the cloth, and after several wires have been woven in, the first one is withdrawn. As it is removed it cuts all the warp threads looped over it, so that they project in the form of tufts of fibres.

The pile wires vary in shape according to whether they are to be inserted and withdrawn mechanically, or by hand. The length of pile depends on the depth of the wire used, and the most common sizes are 1.5 to 3 mm deep. The number of wires retained in the cloth during weaving is usually not less than four.

The pile warp takes up much more rapidly than the ground warp during weaving, due to the looping over the wires, and it is made much longer. By folding velvet cloth across from selvedge to selvedge, it is possible to see rows of tufts where the wires have been inserted.

Diagram 35. Velvet. A = pile warp B = wire C = ground warp D = weft.

The yarns used for velvet are cotton, or linen for the ground warp, with spun silk, reeled silk, rayons or synthetic yarns for the pile and weft. Other velvets include nylon, rayon, or other man-made yarns for ground, pile and weft. The velvets with man-made pile are easily cleaned and stains such as ink and wine can be removed with blotting paper. Velvets are subjected to various finishing processes.

PILE FABRICS 121

GENOA VELVET

This type of velvet has a multi-coloured pile on a satin ground. The pile may be cut or uncut and the yarns used are natural and man-made.

TERRY VELVET

A velvet woven with the aid of a loose reed as in towelling, so that loops are formed for the pile.

UTRECHT VELVET

In weaving ordinary velvet, all the pile threads are over each wire, thus forming horizontal rows of tufts. In Utrecht velvet half the pile threads are over each wire, so that the tufts are arranged alternately and the foundation cloth is more uniformly covered by this method. All the pile warp may be brought from the same beam, but separate slackening bars are used for the odd and even pile threads to assist in the weaving.

The pile threads are firmly stitched into the foundation and the ground weave is 2 and 2 warp rib, but the pile and ground weaves together give a 2 and 2 hopsack foundation. The pile warp may be mohair, silk, or man-made yarns and the ground warp and weft in cotton.

Utrecht velvet is largely used for upholstery fabrics, as it is a hard-wearing cloth.

FIGURED VELVET

This type of cloth has a plain or flat background and a pile figure. The ground weave is usually a closely set rib to obscure the pile warp woven in at the back of the cloth. The pile is brought up in the areas where it is required, and may be cut, uncut, or a combination of the two.

Figured velvet is hard-wearing and is used for upholstery or curtaining. It is rather expensive in good quality yarns, such as cotton for the ground and silk for the pile. Various natural and man-made yarns are employed for figured velvet, either alone or in combination with each other.

Sanderson & Sons Ltd. Plate 11. Figured Velvet. Width of material 52"/54". Repeat 16" length, 13½" width. Woven in 60% rayon, 40% cotton, with cut and uncut pile.

PILE FABRICS

VELOUR

Velour is the French word for velvet. This fabric has a warp pile. The cloth is densely covered in short mercerised cotton pile with a plain backing. The depth of pile can be seen by removing a weft pick from the edge of the cloth, as this has the pile tufts attached to it.

Velour is used as an upholstery fabric for large chairs and settees, as it is a hard-wearing cloth in a good quality. It is also suitable as a curtain fabric in a finer yarn size than is used for upholstery.

Originally, velours were woollen cloths, felted and raised and used for hats and some suiting.

VELVETEEN

In this type of cloth the weft yarns form the pile. The ground picks interweave with the warp to form a firm foundation, while the pile picks float loosely on the surface. These are eventually cut to form the pile. After the cloth has been woven the surface is densely crowded with floating picks of weft, and on the underside a plain rib or fine twill weave is shown.

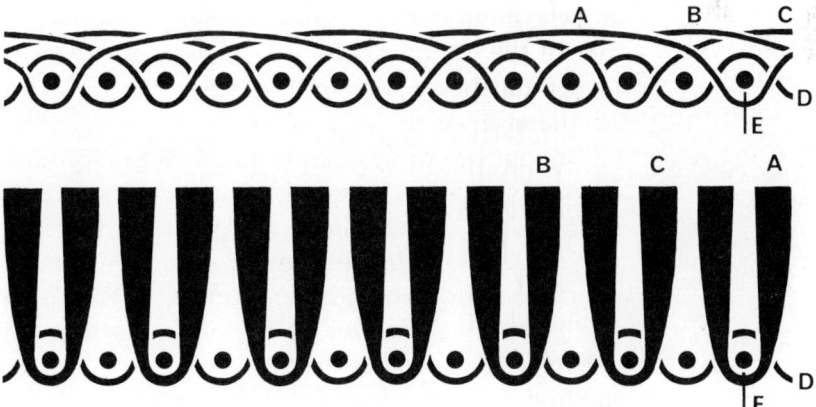

Diagram 36. Velveteen. A = 1st weft B = 2nd weft C = 3rd weft D = ground weft E = warp.

Prior to the weft floats being cut, the pile side of the cloth has a stiffening paste applied. The fabric is dried and a further coating of paste is applied to the underside, followed by drying. The application of these substances fixes the pile weft threads so that they will not be pulled out during the cutting process.

A length of the prepared cloth is stretched very tightly in a frame, and a finely pointed guide passes under the pile floats in line with its movement and brings them into the correct position for cutting. A knife cuts the pile threads on the guide, travelling with it in the direction of the length of cloth until all the pile threads are cut.

After the cutting operation the stiffening substances are removed by scouring and the cloth is then dried and passes through the finishing processes.

Velveteen is woven from 100 per cent cotton, or cotton and rayons, and used for furnishing fabrics.

FIGURED VELVETEEN

This class of fabric is composed of a flat background, and the figure carried out in pile. They can be of an elaborate nature with most of the surface covered in the pile figure, and the ground used chiefly to separate the parts of the figure. The construction of the pile follows that of a simple velveteen, but in the ground portion the pile weft must be obscured from the face of the fabric.

There are three main methods of producing figured velveteen, and in one the pile picks not required on the face of the cloth for the development of the figure are interwoven on the underside in the same manner as on the face. In the second method the surplus weft is allowed to float freely on the back of the ground portion and after the cutting operation is brushed away as waste material. Another method is to interweave the unwanted pile picks with a series of extra ends so that a tissue is formed on the back of the cloth. This imperfect cloth is pulled away, leaving

PILE FABRICS

the ground clear of all surplus pile weft, after the cutting operation.

Figured velveteens are woven in cotton and man-made yarns in rather elaborate designs and used for curtain and upholstery materials.

CORDUROY

These fabrics are made in a variety of qualities with cords running the length of the material. In the finer classes of corduroy, which are used for children's clothing and dress fabrics, fine yarns with a plain background weave are used. For men's and women's clothing, heavier and stronger corduroy is produced with a twill foundation weave. The width of the cords varies and can be from 6 to 20 mm.

The pile is formed by the weft floats which are cut along the length of the cords, making the fibres project in tufts between the vertical lines of ground weave. Corduroy is woven usually in cotton, but can have a percentage of man-made yarns included. The finishing processes are the same as for velveteen.

NEEDLECORD

A dress fabric with very fine cords in the length of the fabric. The pile is very short and easily crushed. It is usually woven in cotton in a fine yarn size.

DESIGNS ON PLAIN WARP PILE FABRICS

In this method, plain warp pile cloth is passed through engraved rollers, and a design is imprinted on to the surface. The pile, where the figure is required, is pressed down and the ground section left with the pile standing vertically from the foundation. The ground is then partially cut away, and the flattened pile is raised by brushing, so that a distinct pile figure is formed against a flat background.

PLUSH

There are two classes of plush; one is formed by the weft,

and woven in the same manner as velveteen, and the other is formed with the aid of wires, as in velvet weaving. The pile in warp and weft plush is very much longer than the pile in other fabrics.

Weft plush is formed with longer floats on the surface of the cloth than are needed for velveteen. The pile in these fabrics may be produced from worsted, wool, mohair, rayons, synthetic or cotton yarns, and is from 6 to 25 mm in depth. The ground is usually woven in cotton. This type of plush can be printed on the surface, in imitation of leopard skin and ocelot. The fabric can closely imitate fur and is often referred to as fur fabric.

Warp plush is woven with half the pile threads over each wire, with the tufts arranged alternately giving a more uniform covering to the foundation cloth. The pile is of a greater length than usual. The yarns may be cotton for the ground and weft, with silk or man-made for the pile.

The two types of plushes are produced for coating and upholstery, as they are durable, and clean well.

MOQUETTE

There are three types of moquette, one with a cut pile, another with uncut pile, and the third combining both cut and uncut in one fabric.

In a cut pile moquette, there are two methods of production. With the first method, the cloth is produced by weaving two fabrics at once, with a space between them, and each fabric has its own warp and weft. The pile is formed by warp threads interlacing, alternately, with the picks of both fabrics. The length of pile is regulated by the distance that the fabrics are apart. Both cloths are woven with the pile threads extending between them, and a knife, fixed on a carriage, travelling to and fro on a rail, cuts the pile. The two cloths are woven in this manner, with the pile on the upper side of the bottom cloth, and the pile on the lower side of the top cloth. Separate take up rollers are used for the cloths, and they are wound on to two cloth rollers after the cutting process.

The other method of weaving cut moquette is by employing the same method of production as velvet. The wires are inserted in place of a pick in the weft, and as they are withdrawn, they cut the pile threads that are looped over them.

For uncut moquette, the fabric is woven as for velvet, with two warps, one for the ground and one for the pile. The pile is formed with wires inserted in place of a weft pick, where only the warp threads that have to form the pile are raised. Unlike the wires used for velvet, those employed for uncut moquette do not have a cutting blade. When they are removed from the cloth, they leave the pile in the form of loops. The cloth is very firm and regular and may be woven with stripes, checks, spots, or in more elaborate designs. Extra warp is added to give more colours.

In one type of uncut moquette, only half the pile threads are over each wire, so that the loops are arranged alternately. In plain pile structures the foundation weave is usually plain, rib or hopsack.

Moquettes that have a combination of cut and uncut pile are woven with the aid of wires. On one pick the pile warp is cut with the removal of a cutting wire, and on another pick the wire is without a cutting blade and leaves the pile in loops when it is removed. Thus the fabric is formed with some portions of the design with cut pile and other parts with loops of uncut pile.

In both cut and uncut moquette, the pile may be in worsted, mohair, nylon, or blends of natural and man-made yarns, with cotton for the ground. In some cheaper qualities of fabrics all cotton may be used for the pile and ground, but this does not give a good wearing cloth and the pile tends to flatten easily. Fabrics with a worsted, mohair or nylon pile give hard-wearing cloths and a pile that springs back into its original position after being flattened.

Moquette is one of the best upholstery fabrics and is used for public transport, theatre, and cinema seating. It is not so popular now for domestic use.

TOWELLING

Turkish towelling is a class of warp pile, also referred to as 'Terry' pile. Two series of warp threads are placed on separate beams, one for the ground threads and one for the pile threads. Only one kind of weft may be used and this interweaves with the ground threads, giving a foundation texture from which the pile threads project in the form of loops.

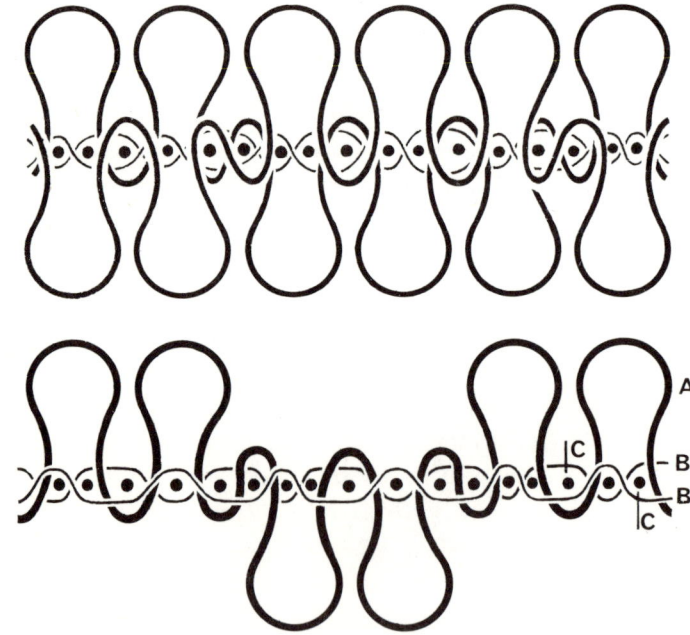

Diagram 37. Towelling. A = pile warp B = ground warp C = weft

The loops are not produced with wires, but by a special reed motion and warp easing arrangement. This system enables the loops to be formed on the upper or lower surface of the cloth, or on both sides of the fabric.

The formation of the loops is achieved by causing two succeeding picks of weft to be left a short distance from the fell of the cloth, and then beaten up with the third pick after it has been inserted. The first two picks are in the same shed made by tightly held ground threads, and the pile threads change from one side of the cloth to the other between the first two picks, and are gripped at this point between the two. As the three picks are beaten up, the slack pile warp threads are drawn forward and form two horizontal rows of loops, one projecting from the face and the other from the back of the cloth. In some cases four, five, or even six picks are inserted in making each horizontal row of loops.

Turkish towelling may be made in either linen or cotton yarns in ground and pile or in cotton for both, and apart from towels has many uses as a domestic fabric in beach wear, mats, and bathing gowns.

In terry ornamentation, coloured pile threads are introduced to form stripes, but if the loops are formed on both sides of the cloth, one side may be of a different colour to the other. The change of colour from back to front of the cloth can also be seen with check designs.

CHAPTER 12

Net Curtaining and Vision Nets

Leno and Gauze; Marquisette; Madras Muslin; Gossamer; Vision Nets; Swivel and Lappet Weaving

LENO AND GAUZE

The net curtaining that is produced today is woven in cotton, nylon, Terylene and other synthetic yarns. Woven fabrics are normally made up of vertical and horizontal threads, placed parallel to one another. For net curtaining a different method is used, in that the warp threads are not parallel, but cross over each other to form the weave.

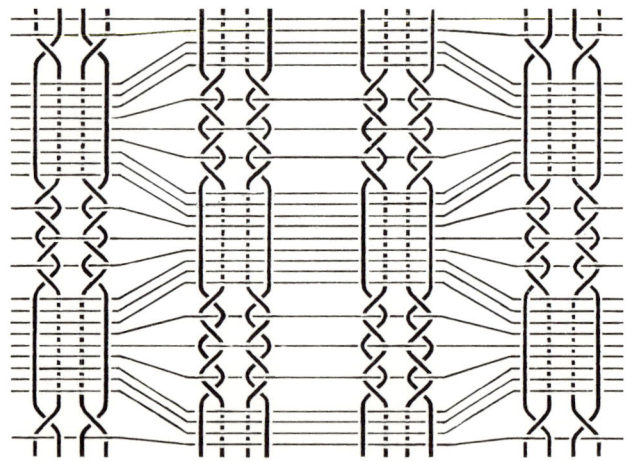

Diagram 38. Net Curtaining in Leno Weave.

These fabrics are termed leno and gauze and come under the heading of Cross Weaving. In leno and gauze weaving, certain ends (crossing ends) are passed from side to side of the normal straight ends (standard ends) and are bound in that position by the weft. The term gauze is applied to fabrics, where the crossing ends pass from one side to the other of the standard ends on succeeding picks. The term leno is used for fabrics in which one or more picks are inserted between succeeding movements of the crossing ends.

The yarns used for the crossing ends must be of good quality, uniform in thickness and smooth, with as little loose fibre on the surface as possible. The crossing of the threads may be made from a single warp, brought from one beam, but in the majority there is such a difference in the take up that two beams are used, one for the crossing ends and one for the standard ends.

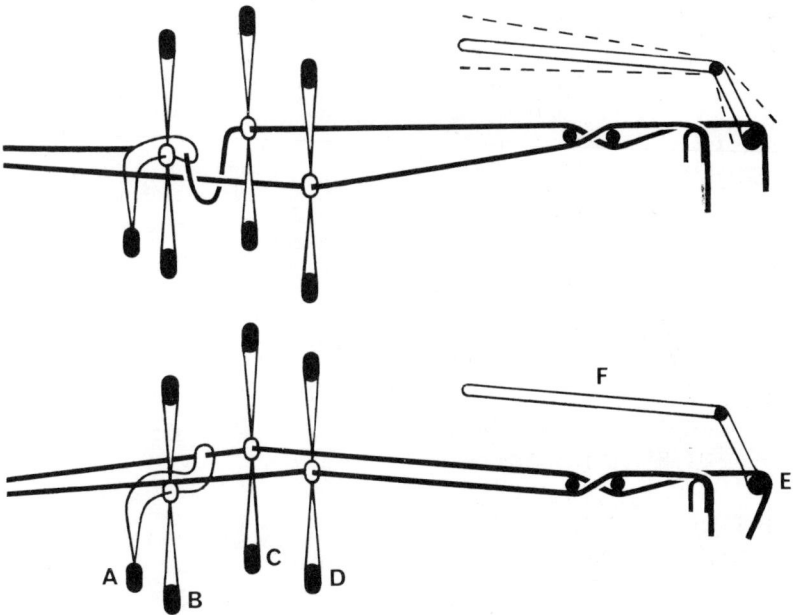

Diagram 39. Leno and Gauze Weaving. A = doup B = front crossing heald C = back crossing heald D = standard end E = easer bar F = lever.

Leno and gauze fabrics are woven mainly on the dobby loom and occasionally on the Jacquard, with additions made to the ordinary shedding arrangements. These consist of a half heald (termed doup or slip), connected to an ordinary heald or harness and a slackening or easing motion. In some types of fabrics, a shaker motion assists in raising the standard ends to the centre of the shed and returns them to the bottom, when the crossing has been effected.

The crossing ends pass through the loops of the half heald, which allows the maximum movement of the end. As the cloth is woven the crossing ends in their doups pass from one side of the standard ends and are held in position by a pick of weft, and then pass to the other side and are held in place by another pick of weft. This process is repeated throughout the cloth, with the crossing ends passing from one side to the other of the standard ends, thus producing an open, perforated fabric.

MARQUISETTE

An open, loose, curtaining fabric, woven from cotton (usually mercerised), silk and wool, but the majority are produced in synthetic yarns. It is woven in white, in a leno construction, and may then be dyed or printed. Spots or other designs are sometimes woven in extra white, or in various colours, on a swivel loom (page 134). Some marquisettes are not woven, but are produced on a simple type of lace machine. Swiss Marquisettes are manufactured in this way.

MADRAS MUSLIN

A fine and open gauze foundation is used, and figured textures and designs are produced by using thick, soft spun, extra weft threads. For curtains, an open ground effect is produced with white or light colours, and there may be additional figuring yarns. In the better qualities of fabric, two or more different colours in extra weft are introduced for the formation of the figure. For cheaper materials, only one figuring weft is used, and to imitate a multi-coloured

NET CURTAINING AND VISION NETS 133

effect, the fabrics have additional colours printed on to the figure.

GOSSAMER

A very soft, flexible, silk or synthetic gauze fabric, woven with one end crossing one end. The yarns employed are in a fine denier, with about 44 ends and 80 picks per inch. Gossamer is principally used for veilings.

VISION NETS

These fabrics are coarse in comparison with the Terylene or nylon nets. They are manufactured for curtaining, room dividers, or screens, and are very effective as wall coverings. The colours are usually natural, white, cream, and pale brown, in yarns of cotton, linen, wool and man-made. Fancy yarns are also employed to give added texture to the fabric.

Some vision nets are rather stiff to handle, and do not drape well as curtaining, and are better for screens or room dividers.

The weaving of vision nets is carried out by various methods. In one type of cloth the warp is spaced out, so that a solid stripe is woven in plain or tabby weave: this may be 6 to 25 mm in width, followed by a space in the warp, thus producing a weft stripe. The stripes are made by entering the warp in its normal manner of one or two ends per dent in the reed, followed by a number of empty dents, and a stripe of warp is entered again. This is repeated across the width of the cloth. There are few ends and picks in vision nets, with both warp and weft spacing to produce an open fabric.

For other vision nets, the leno weave is employed. In one type of cloth, the ends per inch are very sparse, with two ends of leno at a distance of 6 to 20 mm ($\frac{1}{4}$ to $\frac{3}{4}$ in.) apart. The weft is also very spaced out with about 7 picks per inch in some fabrics. The leno weave can be combined with plain weave, and a warp stripe may consist of two ends of leno at each side of the plain woven stripe. The leno holds the first and last two ends at the edges of the tabby stripe in position, so that they will not slip or move.

SWIVEL AND LAPPET WEAVING

These fabrics are woven with a plain or gauze weave, and have added decoration in the form of spots, stripes, small isolated flowers, etc. The cloth has the appearance of the designs having been carried out in embroidery. The figure or design is woven on the underside of the plain or gauze cloth, and in swivel weaving is made from extra weft and in lappet weaving from extra warp.

In swivel weaving a number of small shuttles produce the design in extra weft, working in conjunction with a shuttle producing the ground weft with the warp. The swivel arrangement allows the figure to be woven with the least possible wastage of extra yarn. Each figure or part of a figure in a horizontal line of cloth is formed by a separate shuttle. The extra weft is only introduced where it is required, with little or no yarn extending between the figures on the reverse side of the cloth.

The swivel mechanism makes the Jacquard loom it is attached to much more complex and reduces speed and output. After inserting and beating up each ground pick of weft, those ends, under which the swivel weft has to pass, are raised, and the frame carrying the swivel shuttles is lowered into the shed opening. The shuttles move from one holder to another through the shed that has been formed for them, and insert a separate pick of weft. The weaving of the ground is thrown out of action to allow this to happen. When the swivel weft has been inserted, the frame carrying the shuttles is raised out of the way and the swivel picks are beaten up.

In lappet weaving, the warp thread that has to form the figure is on a separate beam, with the ground warp on another beam. The lappet thread comes from the beam and passes through lappet needles, placed in front of the ordinary reed. The mechanism enables the lappet needles to pass through the warp, whilst the shuttle binds in the lappet warp and weaves the ground cloth. The lappet threads are placed horizontally and are bound to the ground by a weft pick at the point where it returns for the next line. When the lappet thread is not required, the needles move in a vertical direction to the point where they have to be raised again to give the required design.

If the design is of separate motifs, the figuring thread travels from one motif to the other and is cut away after the cloth has been woven, leaving each motif detached from the other.

Swivel and lappet fabrics are used for dress materials, fine curtaining, and bed covers, and are woven in natural and man-made yarns.

CHAPTER 13

Plastics and Non-woven Fabrics

Plastic (P.V.C.) Sheeting; Laminated Plastics; Plastic Coated Yarns; Plastic Filament; Bonded Fabrics; Spun Bonds; Felt; Paper Fabrics

The development of plastics has brought a great change economically to various trades and to domestic living. It is sometimes impossible to use fabrics made of natural or man-made yarns due to excessive heat, light, damp, and soiling conditions. To fill this void many plastic fabrics have been developed which withstand those conditions. Plastics in various forms find a wide outlet in the furniture, clothing, shoe, handbag and suitcase trades and in many other industries.

PLASTIC (P.V.C.) SHEETING

The thermo-plastic material P.V.C. (polyvinyl chloride) is used for many purposes. In one form the plastic material may be opaque or translucent. The sheetings are made in different colours and can be plain, embossed, or have a pattern printed on the surface. The colours of the actual material are fast to washing, but cannot take strong sunlight, and the printed design will often be removed by excessive steam or water. It has a high resistance to the normal conditions of damp and water and is ideal for bathroom or kitchen curtains. For raincoats the material is plain and can have stitched or welded seams. Other articles manufactured from plastic sheeting include umbrellas and table-cloths, and it has uses in industrial and electrical purposes.

Another form of material is obtained by coating natural or man-made woven or knitted fabrics with P.V.C. for use as raincoats and bags. The fabrics may be in stripes, spots, checks, etc., or it can be woven plain and printed with a design, prior to the P.V.C. application.

A P.V.C. coated nylon material has been developed to replace a cotton canvas for tents, as it is tough, waterproof, non-absorbent and tear resistant. The seams can also be stitched or welded.

In upholstery plastics a great number of varieties and qualities are produced. The vinyl coating is applied to woven and knitted base fabrics, or fused to a non-woven nylon backing. For the woven base fabrics, a fine cotton yarn is used and the weaves employed are usually plain or twill. The most popular upholstery material at present is an expanded vinyl, produced like a sandwich, with a core of expanded vinyl between a top layer of P.V.C. and a bottom layer of base fabric.

These materials have good stretch properties, especially on a knitted cloth base. They are easy to apply and can be cleaned with soap and water or proprietary cleaners, specifically made for vinyl upholstery. The regular use of strong detergents will damage vinyl materials. The surface is embossed in imitation of leather or hide. In some types the surface is printed and embossed with a pattern or an imitation weave.

Another kind of upholstery fabric has been developed with an acrylic fabric surface and an expanded P.V.C. backing. This gives an upholstery cloth that is easy to clean on the woven side due to the P.V.C. backing.

Vinyl-coated woven and knitted fabrics are also employed for shoe uppers, shoe linings and slippers. Handbags, travel goods, suitcase coverings, notecases, bus and train wall coverings are also produced in these fabrics.

LAMINATED PLASTICS

In this type of plastic, two or more layers of a textile fabric are fused together by a plastic interlining. They are subjected to heat and pressure, and great toughness and strength are obtained in this way.

Laminated plastics are also manufactured using paper or wood as the base materials. Where heat resistant plastics are required it is customary to use asbestos fibres. Although the plastic application protects the fibre base from heat action, the asbestos has a greater resistance than natural or man-made yarns.

Other types of laminated fabrics are produced for belts, collars, etc. These are semi-stiff and are composed of an interlining of cotton and acetate rayon, fused to two outer layers of fabric. The belts and collars are made up and immersed in a solvent so that the acetate rayon is softened. The application of heat and pressure bonds the three layers of fabric together.

PLASTIC COATED YARNS

These yarns are produced by subjecting a natural or man-made yarn to a bath containing a plastic solution. A fine covering of the compound adheres to the yarn, which is then dried. This process is repeated several times until the required degree of thickness is obtained without losing flexibility. Plastic coated yarns are used for upholstery and lamp-shade fabrics.

PLASTIC FILAMENT

Material is woven from plastic filament yarn, which is wholly synthetic, and is extruded in the same manner as man-made filaments. The weaves are twill, plain, herringbone and distorted weft and the fabrics are used for speaker covers on radio and television sets.

BONDED FABRICS

A lace or woven fabric is bonded on to a tricot acetate backing and has a covering of vinyl. These fabrics can be used for raincoats as they are waterproof and can be made up by stitching or welding the material without the need for a lining.

SPUN BONDS

These non-woven materials are made from a wide range of fibres such as cotton, rayon, acetate, viscose, nylon, etc. A mass of fibres in staple form are opened up and cleaned, so that they become fluffy and light. The fibres may be used alone or blended with each other, but an equal percentage of each fibre must be used in the blended variety.

A web is formed by a carding machine, and as the webs emerge from the carders they are laid one on top of the other. There are several methods of making a web construction; either by cross laying, in which the webs are laid in different directions, or by parallel laying so that all the fibres are in one direction. Another method is the combination of cross laying and parallel laying, and this gives equal strength lengthways and widthways, but requires three or four layers of web.

The web constructions are bonded together to form the fabric either by heat and pressure, or with an adhesive or bonding agent. Some bonded fabrics have an adhesive sprayed on, or are totally immersed in the bonding solution. The fabrics can be printed by the usual methods and the dyes used are those most suitable for the fibres forming the web.

These fabrics when made up into garments may be washed up to 12 times and are quick drying. They have a good resistance to creasing and can be pleated and sewn. The main uses of spun bond fabrics is in disposable dresses, hand towels, tablecloths, napkins, etc.

FELT

The wool fibres from a woollen carding machine are arranged layer upon layer until the desired thickness and the width of the carding machine is built up. Simultaneously, the required length is built up in the same manner. The fibres are then milled or beaten whilst wet with warm soapy water, so that they become interlocked and matted.

Felt is used for a variety of purposes, as a background or as collage work in embroidery, for soft toys, hats and for covering tables.

PAPER FABRIC

This fabric consists of two layers of soft, tissue-like paper and between them is a layer of cotton scrim. The fabric is chemically treated to make it fire resistant and is used for dresses. It is easy to sew and hems are not necessary, the hem line is usually left so that they can be trimmed by the customer to the length required.

Various methods have been tried in printing designs on the paper fabrics, including roller printing and automatic screen printing. It was found that hand screen printing was the most suitable for the fluorescent colours the material required.

CHAPTER 14

Finishing of Fabrics

Desizing; Scouring; Bleaching; Stoving; Mercerisation; Raising; Nap Finish; Dress Face Finish; Cropping or Shearing; Brushing and Steaming; Singeing; Beetling; Crabbing; Milling; Stentering; Waterproofing; Velanizing; Crease Resistance; Mothproofing; Rot proofing; Flame Resistance; Sanforizing; Calendering; Schreiner Calender; Moiré Fabrics; Embossing; Velvet and Velveteen Finishing; Pilling

DESIZING

To facilitate weaving, and prevent the warp threads from fraying or breaking with the friction of the healds and reed, they are coated with size. This dressing is usually a modified starch with fats, waxes, or other substances, and for continuous filament polyester yarns a vinyl co-polymer size is employed.

The removal of these sizes is carried out by the method most suited to the type of size used. For a starch size it is treated with a malt extract solution and washed thoroughly. A gelatine size is removed by boiling in a soap solution, followed by washing. Linseed oil, which is frequently applied to viscose yarns, is treated with soap and sodium carbonate (soda ash), followed by washing. In other highly oxidised sizes the cloth is treated with trichloroethylene before scouring. With other substances an appropriate solvent is used and some synthetic resins are removed by dilute acids.

SCOURING

The natural impurities which are soluble must be removed from the fibres by a special operation called scouring. All fibres have their own scouring operation, and this varies from fibre to fibre. Scouring is a preliminary to bleaching and the more efficient the scouring, the less bleaching required. It is also a distinct operation where bleaching is not required.

The scouring agents employed are usually soap or synthetic detergents with or without the addition of an alkali; the latter plays an important part where the removal of fats is required. These agents are selected according to the type of fibre to be purified.

BLEACHING

The bleaching of cloth or yarn follows scouring and washing. The type of fibre determines the bleaching agent to be used, and those most usually employed are hypochlorites, hydrogen peroxide and other peroxides, potassium permanganate, and reducing agents, hydrosulphites, bisulphites, etc.

A peroxide bleach is used economically and successfully on lightweight fabrics, knitted garments, and in some instances on damasks, as the vat dye used for the yarns is not affected by the peroxide.

STOVING

Blankets, knitted woollens, white flannels and other woollen goods are bleached by stoving. The stove is a small brick or wooden building and the goods are suspended from hooks, rollers, etc. inside. The materials are thoroughly scoured, soaped and blued and the water is extracted. Sulphur is then burned and forms sulphur dioxide, and, on contact with the damp materials, forms sulphurous acid. The duration of stoving depends on the type of goods and the amount of bleaching required. The process may take a few hours or all day. At the termination of stoving, the goods are removed, put in the open air and finally dried.

MERCERISATION

A process named after Mercer, who originated it by experimenting with caustic soda on natural fibres and found

FINISHING OF FABRICS

that they could be made permanently lustrous by this process. The imparting of this lustre to cotton may be carried out in the yarn or cloth stage.

Yarn is mercerised in hanks in automatic machines, and usually in the unscoured state. The hanks are rotated continuously over two steel rollers and come in contact with a rubber-covered roller, which squeezes the surplus liquor out and ensures good penetration. The direction of rotating is reversed several times and the hanks are relaxed and stretched while wet with the caustic liquor. The whole operation, including washing, is usually completed in about two minutes.

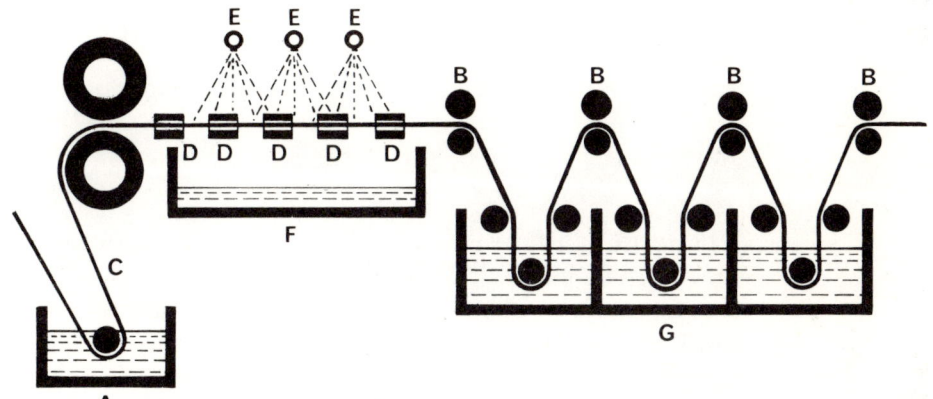

Diagram 40. Mercerising of Fabrics. A = caustic soda roller B = squeezing rollers C = fabric D = stenter clips E = hot water sprays F = rinsing water G = washing unit.
The fabric (C) is padded with caustic soda in the pad (A) and passes between two rollers into the clip stenter (D) where it is stretched and rinsed with hot water (E). From the stenter unit it passes through the squeezing rollers (B) and is washed and rinsed at (G).

In fabric form the process is continuous and the cloth is passed through caustic soda liquor with the surplus liquor squeezed out in a pad under very heavy pressure. The cloth is allowed to relax for a short distance of its travel, while being pulled out to approximately its original dimensions by stenter clips. It is then washed thoroughly, and hot and cold water rinses are applied to remove as much of the alkali as possible. This is followed by further processing as required.

Mercerised linen does not achieve such a high lustre as cotton, but the process increases the capacity of linen to absorb dye. Unions of cotton and rayon can be treated with caustic potash to improve the appearance of the cotton.

RAISING

To impart a hairy surface to a cloth, the fibres are lifted out of the yarns by raising. This is achieved either with small wires on rollers, or with teazles arranged horizontally on a large revolving cylinder. The cloth is stretched or tensioned in the machine and the raising rollers or cylinders rotate rapidly in the opposite direction to the cloth.

The wires or teazles pluck the cloth, drawing out the fibres to the surface of the cloth.

NAP FINISH

The term nap finish is used for fabrics with a fibrous surface, in which the fibres are raised to stand vertically from the foundation of the cloth. The surface may be rubbed into the form of small curls or nubs, or can be left as a rough hairy surface.

DRESS FACE FINISH

A finish applied to heavily felted and raised woollen cloth. The short fibres are laid lengthways on the fabric and completely conceal the weave. The texture is dense and fibrous, achieved by the felting process, and the fibres are combed in one direction by the raising process. A lustre is imparted to the surface by boiling the fabric.

CROPPING OR SHEARING

The shearing machine cuts the irregular pile of fibres level after they have been raised. The cutting part of the machine consists of a horizontal fixed blade, and against it

FINISHING OF FABRICS

revolves a cylinder, containing a number of spiral cutting blades. The fabric passes over an adjustable bed-plate and the closeness of the bed-plate to the knives gives the amount or height of cutting.

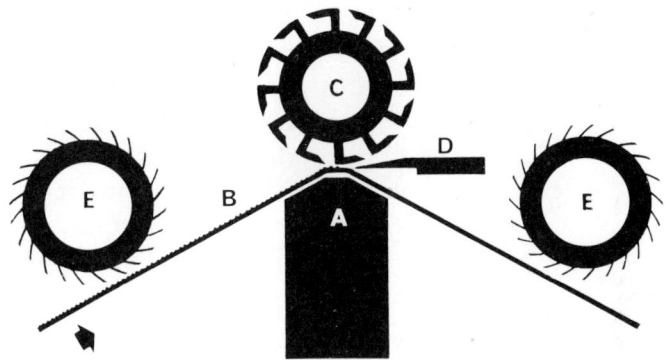

Diagram 41. Shearing. A = table B = fabric C = rotating blades D = knife E = brushes.
The surface of the fabric (B) is brushed by the first roller (E) and passes over the shearing table (A) to be subjected to the cutting action of blades (C). The shearing knife (D) cuts any surplus fluff and brush (E) removes it from the fabric.

A pattern may be produced on a pile fabric by replacing the bed-plate with an embossed roller, which rotates with the cloth, and the unwanted pile is cut away. Mechanical shearing or cropping is used to give well-defined weaves that have become obscured by other finishing processes.

BRUSHING AND STEAMING

These processes are applied to various cloths, depending on the type of finish required. It is used for woollen materials in a double process, both before and after cutting. The steam softens the fibres, and they are then more easily brushed. The cutting process can be followed by further brushing, which removes all loose fibres. The effect of brushing and steaming is to impart a brightening of the surface to the cloth.

Steaming is also used to soften the fibres and give a

lustre to a fabric, by subjecting it to steam for about 10 minutes and afterwards cooling with air. The cloth may then have a light calender or press.

SINGEING

This operation is necessary for smooth and highly lustrous fabrics, and this process particularly applies to good quality linen and cotton cloths. The object of singeing is to remove all loose fibres or fluff from the surface of the cloth. This may be carried out by passing the cloth over two very hot plates, with the heat kept constant, otherwise the singeing will not be uniform and patchiness will result.

Gas singeing gives the best results when the cloth is passed rapidly over two or three sets of burners, at about 146 to 180 metres per minute. Before reaching the first burner, a revolving brush raises all the loose fibres, and after singeing the cloth passes through water troughs in order to extinguish any sparks.

This process is used for cotton, linen, fine woollen and worsted yarns or cloth, to give them a lustre and remove all surplus fibres. The cloth and yarn is afterwards termed gassed yarn or gassed cloth.

BEETLING

Beetling is a finishing process, particularly used for finishing linen cloth, and is used to close up the spaces between the threads, to flatten them and impart a lustre. In one method the roll of cloth is subjected to a rapid succession of blows applied across its surface by wooden hammers with round ends. The rounded ends limit the contact with the cloth to a small area, and the exterior roll receives more treatment than the interior. Batches of cloth are re-wound, to allow both sides of the cloth to be treated.

In another method, the fabric is passed through a pair of steel rollers, which exert great pressure and flatten the fabric.

FINISHING OF FABRICS

CRABBING

To ensure that the warp and weft will not slip on one another and will remain in a permanent position, worsted fabrics are subjected to crabbing. The machine used for this has three units, each one has a trough with guide rollers and over these are mounted two other rollers. The troughs in the first and second unit contain boiling water, and in the third cold water. The cloth under tension is run through the trough of the first unit, and passes on to be wound on the upper rollers. It then passes through the trough of the second unit, and finally to the third unit containing cold water, which cools and sets the cloth.

The material is wound on to a perforated steaming roller and wrapped in a cotton cloth. Steam at pressure is passed through the roller while it rotates. Once more the material passes through cold water to a second steaming roller where the process is repeated. Crabbing is often carried out on worsteds before scouring takes place.

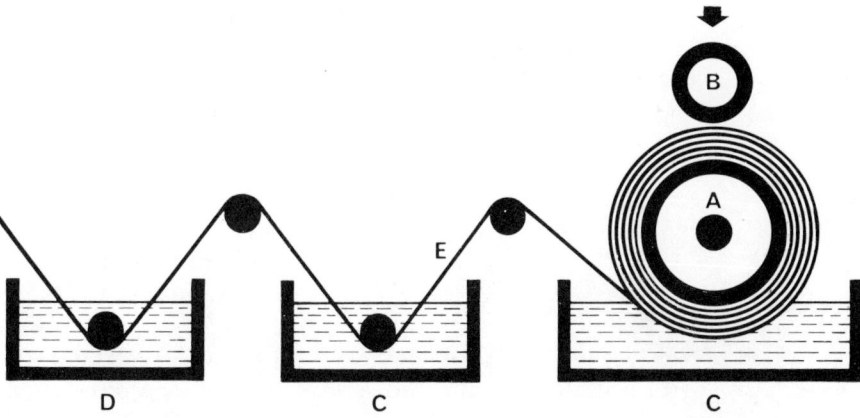

Diagram 42. Crabbing. A = crabbing roller B = pressure roller
C = hot water trough D = cold water trough E = fabric

MILLING

Tweed cloth or other fabrics which require dense textures are passed through a process called milling. This finish felts the yarns together and shrinks the fabric, so that as a garment it will not shrink in the general conditions of wear and weather.

The fabrics in the form of a rope are passed through machines with heavy wooden rollers, fitted with a compressing mechanism. The compression and relaxation of the fabric, in the wet state, causes the felting. Milling may be carried out with hot water only, or with soap and sodium carbonate, or with acid.

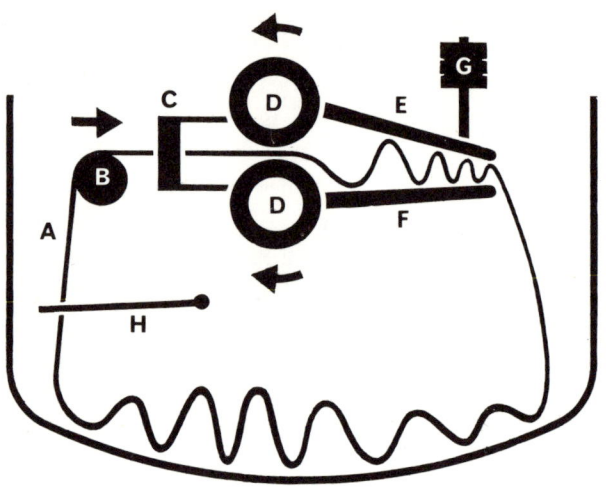

Diagram 43. Rotary Milling Machine. A = fabric B = guide roller C = throat D = rollers E = movable top board F = fixed bottom board G = weights H = draft board.

STENTERING

This finishing process is employed to stretch the fabric to a certain width and is also used in the drying of textiles. The fabric should not be overdried and must contain a certain degree of moisture 'regain' for the best fabrics. This is achieved by storing the fabric in a cool cellar.

There are two types of stenters, those with clips and those using pins, and the fabric is dried in one of these under tension. The stenter with clips holds the fabric at

FINISHING OF FABRICS

each selvedge, and pulls it out to a certain width and length. The clips are on an endless chain and the fabric travels through the machine, while steam pipes blow hot air on to the moist cloth. The fabric is gradually dried, and the speed through the stenter is regulated according to the type of cloth being dried. A mild combined stentering and calendering is frequently given to rayons, and other light fabrics.

In the second method, the cloth is held by pins instead of clips. The fabric travels backwards and forwards in a heated area, on two or more levels, and is heated for longer than the other method allows. For delicate fabrics a winding-on arrangement is used to ensure that the cloth will not be damaged. This method is used more for woollen materials.

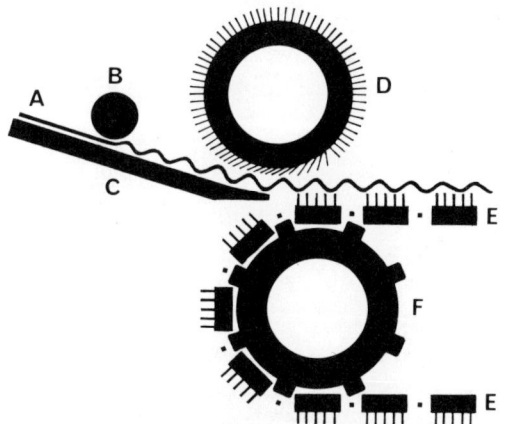

Diagram 44. Stentering. A = fabric B = rubber roller C = brass slide D = pinning brush E = stenter rail F = driving roller. The fabric (A) is pushed on to the brass slide (C) by the rubber roller (B). The fabric is fixed on to the pins of the stenter rail (E) as it is turned by (F) and during drying the ripples on the fabric are taken up.

WATERPROOFING

For waterproofing, all kinds of materials are treated with various solutions suitable to the type of fibres. Nylon fabrics are waterproofed by the use of silicone resins. Cotton, linen and rayon fabrics are treated with aluminium acetate, wax, or various chemicals for ordinary rainproofing. In waterproofing, the surface of the cloth is treated with rubber or gutta percha.

With some fabrics the proofing is in rubber on the inside only, and in a double textured cloth, the rubber is placed between two layers of cloth. For tents and marquees, the cotton cloth is treated with a solution of ammoniacal copper oxide, which, after heat treatment, gives a durable film of gelatinised cellulose.

VELANIZING

This finish imparts to fabrics of cotton, wool, linen, silk or rayons a soft and supple quality that will withstand repeated laundering or dry cleaning. Velan finish combines with the material to create new compounds, which after drying at low temperature and heating, give the fabric an anti-crease and water repellent finish.

CREASE RESISTANCE

Fabrics which would normally crease in general wear can be made wholly or partially crease resistant. This process is carried out after bleaching, dyeing or printing, and can be applied to cotton, linen, rayons, and blended or mixed fabrics.

A melamine formaldehyde resin liquor is prepared and the fabric is impregnated with the liquid, which is followed by drying and baking. The resin passes into the interior of the fibres and remains to give the crease resistant finish. Severe or repeated washing reduces the crease resistant properties.

MOTHPROOFING

Substances such as chlorinated phenols or similar products, may be impregnated in woollen fabrics, to prevent their destruction by the larvae of the clothes moth. The treatment may be carried out during the dyeing or scouring of the

FINISHING OF FABRICS

fabrics with some substances, as they have a direct affinity for the wool fibres.

ROTPROOFING

All natural fibres are subject to attack by mildew or bacteria, and by using suitable chemicals, this can be avoided. Rotting can be prevented in some fabrics by using mineral dyes, or by treating them with copper or chromium salts.

FLAME RESISTANCE

To render fabrics non-inflammable, they may be treated with mineral salts, such as borax, alum, phosphate of soda or a compound of titanium or antimony salts.

Cotton and viscose rayon fabrics can be treated with a compound of phosphorus called tetrakis-hydroxymethyl-phosphonium-chloride, THPC. This compound provides efficient flame protection when it is used in conjunction with a heat-cured resin. Another method of treating cotton is to wet the fabrics in a solution of tetrakis (hydroxymethyl) phosphonium hydroxide, called THPOH. Partial drying takes place, and the fabrics are exposed to ammonia vapour which reacts with the THPOH and forms a permanent polymer inside the fabrics. Care is taken not to dry the materials too much during the treatment, as this reduces the flame resistant quality.

SANFORIZING

This process is applied to cotton or linen garments, so that they will not shrink or stretch during washing or pressing. The Sanforizing machine shrinks the material to the required degree and further shrinkage in the garment cannot afterwards take place.

CALENDERING

To impart a glaze or lustre to cotton fabrics, such as a chintz, the fabrics are calendered. The lengths of cloth are subjected to friction as well as heat and pressure, by passing through alternate steel bowls (rollers), and cloth or paper covered bowls. The steel bowls are heated and, combined with the weight, cause the closing together of the threads of the fabric.

The fabric emerges with a smooth glossy appearance, produced by the combination of moisture in the cloth, heat and pressure. The amount of gloss depends on the type of calender machine employed, the class of fabric being calendered, and the temperature and pressure.

SCHREINER CALENDER

In this machine the steel bowl is engraved with fine diagonal or parallel lines. With this type of calender, the transparent and gossamer sheer synthetic fabrics can be made more opaque. The threads are flattened to fill out the intersections in the fabric without impairing the lustre.

MOIRÉ FABRICS

Moiré is the French word for 'watered' and moiré fabrics resemble a water reflection. The effect is achieved by folding the cloth warp-ways along the middle, and passing it through the calender machine. The material is sprayed with water or entered very damp, and the heat and pressure imprint the markings from one half of the material to the other.

Another method for moiré is to pass the material between two rollers, of which, one, turning at a greater speed than the other, is engraved with a moiré design. This method gives a more even effect to the cloth. Moiré finish is applied to mercerised cotton fabrics, and those made of silk, rayon and synthetic yarns, which are ribbed either lengthways or across the width.

EMBOSSING

An embossed effect may be given to many types of fabrics, including those made from natural and man-made yarns or plastics. The embossing is achieved by passing the fabrics through a calender machine, with an engraved roller, which imprints a design on to them in the form of ridges or texture. In some fabrics, mainly cotton or viscose rayon, the cloth is treated with a synthetic resin, principally urea formaldehyde or melamine formaldehyde. The treated cloth is embossed at a high temperature and cured at an even higher temperature to fix the resin.

VELVET AND VELVETEEN FINISHING

Velvet may be piece dyed, and in a cloth with a cotton ground and viscose rayon pile, it can be dyed in one process. A fabric containing two fibres that do not have an affinity for the same dye, may be cross-dyed. Velvets are brushed to remove all surplus fibres and the pile is sheared; this can be carried out several times for a good quality of cloth.

In velveteen finishing, the cloth is scoured to remove the stiffening substances, after the pile floats have been cut. It is dried and brushed with flat brushes that work across the cloth in opposite directions, followed by further brushing with circular brushes. Superfluous fibre is removed by cropping or singeing, and in a good quality the brushing and singeing may be repeated several times.

In piece dyed velveteen, the cloth is dyed in the ordinary manner, but the ground tends to be a darker shade than the pile. To remedy this, it is passed in contact with a finely engraved roller, which revolves partly immersed in a dye solution. The dye is transferred by the roller to the pile surface of the cloth, and revolving brushes spread the colour evenly. The dye is fixed, the cloth is dried and brushed again to bring up the pile.

PILLING

This is not a finishing process but refers to the formation of numerous scattered small balls or pills, of fibre tufts, which have become entangled and are loosely attached to the surface of a fabric. They are held to the surface with a few fibres and can easily be pulled away. This blemish on the face of material is caused by rubbing, and can spoil the appearance of any knitwear or cloth.

To avoid pilling, the fabric is singed at a late stage in dyeing or finishing. In another method, the surface of the fabric is brushed to give a pile, sheared to a uniform height, and treated so that protruding ends of fibres shrink and are not noticeable on the face of the cloth.

The yarns used in either knitwear or fabrics should be evenly spun, with sufficient twist to prevent the fibre ends from becoming loose and working out to the surface.

CHAPTER 15

The Printing of Textiles

Block Printing; Screen Printing; Machine Screen Printing; Rotary Screen Printing; Duplex Printing; Transfer Printing; Dye Mixing for Printing

The depositing of colour, by various means, on fabrics to obtain coloured patterns or designs, is termed 'textile printing'. The ornamentation of textile fabrics by printing has been known for at least two thousand years. Fragments of cloth decorated with patterns have been found in the Pyramids and Egyptian tombs. The earliest type of printing was from wood blocks and it would appear that China and India were the first countries to develop this craft.

Printed textiles have become a household requisite, combining colour and design to improve the appearance of a room, or a method of introducing colour and pattern to clothing. With printed textiles it is possible to impress on the fabric a naturalistic copy of an object, and a flower may be printed in such a manner that it becomes a faithful representation, whereas in a woven flower, the fine details are not possible, due to the threads having to interlace to form the shapes and lines. Printing has another advantage over weaving in that areas of pure colour are obtainable and are not textured by the weave. A big range of subtle colour changes can be achieved with overprinting, using only a few dyes.

The design for a print is carried out on drawing paper, with paints, inks, coloured tissue paper or by any other media. Particular attention is paid to the repeat of the design, as it must fit evenly into the width of the cloth, but lengthways it is limited only by the size of the equipment used for printing. The repeat may be full drop, which is side by side, with the top of one repeat fitting against the bottom of the next repeat, or half drop, where the top line of one repeat is placed half way down the repeat next to it. A similar repeat is the brick, when the edges of one repeat are placed in the centre of the one above. The half drop is a vertical stripe and the brick a horizontal stripe.

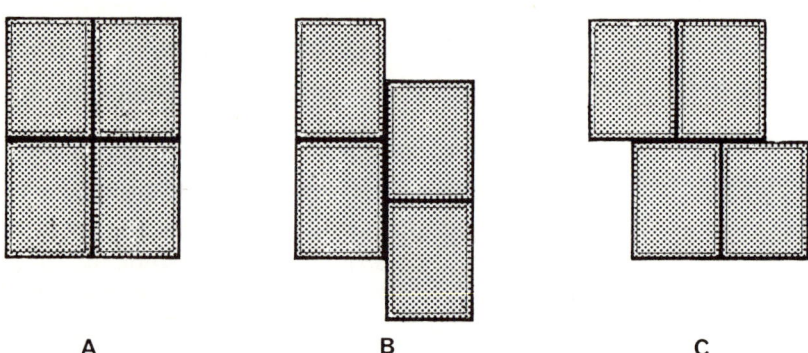

Diagram 45. Design Repeats. A = full drop repeat B = half drop repeat C = brick repeat.

The repeats are most important as a fabric may be marred by the design forming lines, diagonally, horizontally or vertically, when they are unintentional, and spaces or gaps when the design is an overall pattern. The designer decides on the type of repeat which shows the design most favourably, and indicates on the drawing paper how this would be printed. Most designers show one complete repeat and a portion of design around it.

On a furnishing fabric, the repeat must fit into a material width of 48 to 50 ins (122–127 cm), and for dress fabrics, the repeat must be a unit that fits into a material width of 36 in. (91 cm). When the design repeat has been established, there are various methods of printing the colours on to the cloth, but the majority of industrial fabrics are printed by screens.

BLOCK PRINTING

This method is used by craftsmen or small firms for short lengths of material, ties and scarves. The block may be made of sycamore, lime, plane or pear wood, or any other fairly hard, close-grained wood. For large blocks a combination of woods may be glued together, with the sycamore forming the top layer. The grain of each piece runs in a different direction and the block is placed under pressure during the glueing process. By using a tongue and groove method, a stronger and warp-proof block will be made.

The size of the block depends on the width of repeat, but the maximum size is $18 \times 18 \times 2\frac{1}{2}$ in. ($46 \times 46 \times 6$ cm) with a maximum weight of 10 lbs ($4\frac{1}{2}$ Kilos). In some large repeats two blocks may be necessary, each carrying half the repeat. The shape of the block also depends on the design, and it may be either square, oblong, diamond shaped or an irregular form.

In cutting the block the design is traced on to the surface and the areas that do not require to be cut away are tinted with colour. The wood to be removed is cut to a depth of from $\frac{1}{4}$ in. (6 mm) to $\frac{3}{4}$ in. (12 mm) and small pins of brass are driven in at the corners of the block. These pins mark the cloth, and the printer, using them as a guide, can accurately position the block for the next print. To reinforce parts of the design, copper or brass strip is hammered in and felt may be pressed in between these strips for the printing of large areas of colour.

Another method of block making is to coat the wooden block with varnish and dust it with a fine woollen flock of ground felt. This process is repeated until the surface is sufficiently and strongly covered and the flock will not be removed in printing. A block may also be formed without cutting and by building the design up on the flat surface with copper strips, rods or pins of copper or brass. These are hammered in and the surface filed down level, with felt added if required. Another method of block-making consists of a metal cast in the form of the design, which is screwed to the flat surface of the printing block.

The number of blocks required to print a design corresponds to the number of colours in the design. Each block represents one colour and a design containing five colours will require five blocks, each one having on it that part of the design.

The cloth to be printed is extended along the top of the printing table. Dye is brushed across the surface of the sieve (a wooden frame tightly covered with a fine woollen cloth), and the block is evenly coated with dye by pressing it gently on to the sieve. The block is then placed on to the cloth, the back of it tapped with a mallet and the colour transferred to the cloth. After the first impression has been made, the sieve is recharged with dye ready for the next print.

SCREEN PRINTING

This method of printing designs on cloth is derived from the Japanese method of stencilling delicate patterns on fine silk and cotton muslin. In screen printing the stencil plate is replaced by a rectangular metal frame, covered with a fine mesh of nylon or Terylene. The mesh is woven in plain weave and the gauge of mesh is dictated by the thickness of yarn, the ends and picks per inch or centimetre in the construction of the fabric. The gauge of mesh employed for the screen depends on the image to be printed; a fine gauge is used for fine line printing and a coarser gauge for larger areas of colour. The screens are of a standard size and the width of the printing table.

The design is traced from the paper design on to Permatrace, Kodatrace or Ethulon, using an opaque ink or paint. Every colour in the design is traced on to one sheet of Permatrace, as each colour has its own screen. This means that if the paper design is in five colours, five separate tracings will be made. If a design has fine lines or stippling, these are

THE PRINTING OF TEXTILES

traced on to a separate sheet to make an extra screen. The five-colour design would then require six screens.

With the photographic method of developing a screen, it is coated with a photo-emulsion which is allowed to dry thoroughly before the screen is developed. One type of developing table consists of a series of ultra-violet tubes, fitted close together. The Permatrace is placed on a glass surface above the tubes, and the screen, on a movable carriage-way, is moved into position above it. A rail at the side is set with the size of repeat.

The screen is lowered into close contact with the Permatrace. A rubber blanket is put inside the screen and, with a board on top, they are clamped into position. The Permatrace in contact with the sensitised screen acts as a photographic negative and is exposed to the light. The screen is exposed for a period of time for one repeat and moved on to expose the other repeats. All the areas or lines which have been painted on to the Permatrace protect those parts of the screen from the light. This in turn prevents the photo-emulsion from hardening and it can be removed after exposure is completed by washing away in warm water. After the screen has been thoroughly washed, the original mesh is revealed where areas of colour are required on the cloth.

MACHINE SCREEN PRINTING

In this method, the fabric to be printed is on a roll at one end of the printing table. It travels through a series of rollers and comes into contact with a suction pump which removes any loose fibrous material adhering to the surface. From these rollers, the fabric is fed on to a table with a conveyor belt surface. This carries the cloth under the printing screens and automatically stops under each one.

The screens are placed side by side a few inches above the table and are lowered on to the cloth when printing commences. Each screen prints a different colour across the width of the cloth. Two fine blades, called squeegees, are positioned within the framework of the screen, and above them is a channel of dye. A controlled amount of dye is allowed on to the blades from the channel or trough and this is kept constantly supplied with dye.

Before printing commences the conveyor belt stops, the screens are lowered and the squeegee blades move across the screens, taking the colour with them. One blade takes the dye across the screen, pressing the dye through the mesh and on to the fabric. It then moves up out of position and the other blade is lowered to bring the dye back over the screen surface and on to the cloth. The mechanism can be set so that one screen can be printed two, three or four times, depending on the depth of colour required. After the blades have pressed the dye through on to the cloth, the screens rise and the conveyor belt moves the cloth along into position under the next screen. The screens descend and the process is restarted, so that when the fabric reaches the last screen all the colours in the design have been printed.

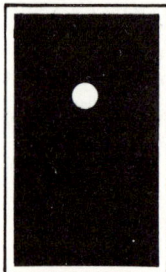   

Diagram 46.   Three Screens and Completed Print.

From the printing table the printed fabric enters a drying chamber placed at the end of the printing table. This is a square chamber with a controlled temperature and the fabric travels through it over a series of rollers and is collected at the other end. The printed cloth is finished by passing through a steaming chamber, acid baths and a scouring unit, and is finally rinsed and dried. These processes fix the dye

THE PRINTING OF TEXTILES 161

and remove the gum used in printing, and the cloth may then have any further finishing processes that are required.

ROTARY SCREEN PRINTING

This method of printing is quick and efficient and allows a greater quantity of material to be printed in a short space of time. The machine employed is rather expensive, but with a greater production of printed cloth it is more economical.

The screen is a seamless electroformed nickel mesh roller and the mesh varies in the number of perforations or holes. The mesh perforations can be round, square, oval, oblong or hexagonal and supplied in either 40, 60, 80 or 100 size mesh. These figures refer to the number of cells or perforations per linear inch. The 40 mesh, being coarser, would be used for printing very heavy fabrics, and the 100 mesh for fine line work and light-weight synthetic fabrics. The most usual circumference size of the roller is $25\frac{1}{4}$ in. (64 cm), but rollers of 19 in. (48 cm) are also employed.

The screen roller is cleansed with chromic acid and thoroughly rinsed, and coated with several applications of emulsion. After each coating, the roller is dried in a special temperature-and-humidity-controlled drying cabinet. The coating of the screens is carried out by standing the roller on end in a machine. A ring carrying the emulsion travels down the roller, depositing an even layer of emulsion on the roller surface. The speed at which the ring travels is dictated by a control panel at the side of the machine.

In order to develop the design on the mesh roller, an inner tube is inserted inside. This is fixed on to a machine and air is pumped into the tube so that it becomes hard and firmly supports the mesh roller. The size of each repeat is marked out on the roller and the Permatrace is wound around it for the first repeat. All other parts are covered to avoid exposure to the light.

The roller revolves beside a mercury vapour light, which has also been masked, except for the width of one repeat, and the Permatrace is exposed for eight to ten minutes. After exposure, the Permatrace is moved to the next repeat and all other parts of the roller and light are masked. This is repeated over the whole length of the roller. Another method of developing is to have the Permatrace film the full width of the roller, and this system avoids any error that might occur in registering each exposed repeat in turn. The air is released from the rubber tube and the roller is taken from the machine.

The roller is developed in cold water and the unexposed emulsion dissolves and is washed out of the mesh. After washing, the roller is cleaned with soft cloths both inside and out and hosed with water. It is then polymerised at a temperature of 180°F in a drying cabinet and trimmed to a specific length, and rings are fitted at each end.

All the rollers are fitted on to the printing machine and adjustments can be made to position the rollers both widthways and lengthways for the repeat. The squeegee is a fine steel blade which is fitted through the centre of the roller and the dye is fed in through a pipe attachment. Each screen roller prints one colour across the width of the cloth, and up to 20 rollers may be used at one time.

The printing table has a conveyor belt surface and the screen rollers are lowered to the cloth surface when printing commences. The cloth to be printed travels from a roller at the end of the printing table over a series of guide rollers and comes into contact with the suction pump which removes loose fibre material. The cloth is fed on to the gummed conveyor belt surface and as the printing roller revolves in close contact with the cloth, the stationary squeegee presses the dye through the screen and prints colour on to the cloth. The pressure of the squeegee can be adjusted to suit each individual roller. The dye is automatically pumped through the tube and a controlled supply of dye flows on to the squeegee.

While the rollers are revolving in one direction, the conveyor belt moves in the opposite. The cloth travels under the

THE PRINTING OF TEXTILES 163

rollers and is printed by each in turn. The speed of the machine and the raising and lowering of the rollers is controlled by a system of dials at the side of the machine. To ensure a clean printing surface, the endless conveyor belt is

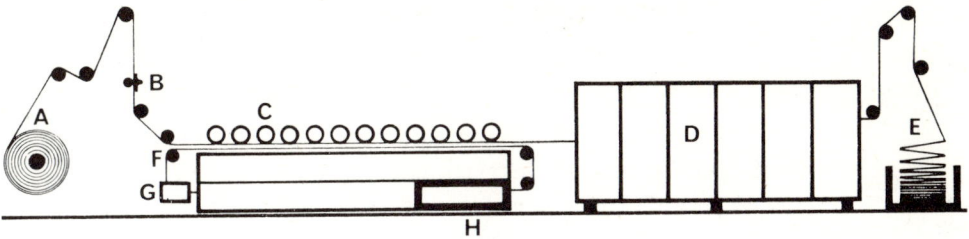

Diagram 47. Rotary Screen Printing. A = fabric to be printed B = cloth feeding device C = screen rollers D = drying chamber E = printed cloth F = conveyor belt blanket G = glueing device H = washing unit

washed as it travels underneath the machine. This removes all the surplus dye and the belt has a fresh coating of gum applied to the surface. After the material has been printed it travels over guide rollers and enters a drying unit placed at the end of the table. The squeegee units and rollers are removed from the machine and cleaned with cold water sprays to remove all traces of dye, and the printed fabric receives the necessary finishing processes.

DUPLEX PRINTING

Duplex printing is employed for heavier fabrics in which the colour penetration is more difficult. Two machines are employed and the fabric is fed through while both machines are run simultaneously, so that both sides of the cloth are printed in one operation. To obtain the correct fit of the design on the face and back of the fabric, careful adjustments and the synchronisation of the machines are necessary.

Further developments in duplex printing are being carried out to print a heavyweight cloth with different designs on each side of the cloth. The fabric has to be of a sufficient thickness to prevent any area of dye seeping through from one side of the cloth to the other. The application of dye has to be strictly controlled to allow enough to be absorbed into the cloth surface giving the necessary penetration to withstand future cleaning processes without a loss of colour.

After printing by duplex the cloth is immediately dried and receives fixation and finishing processes.

TRANSFER PRINTING

This type of printing is an adaptation of the domestic method of ironing embroidery transfers on to a fabric. In transfer printing the design is printed on to paper which is wound on to a roller and placed in position in the machine. The design is the full width of the cloth to be printed. The fabric on another roller is placed opposite the design paper roller. Both rollers turn at an equal speed and the paper and fabric come together in a dry heat press which transfers the design from the paper on to the cloth. All the colours in the design are transferred by the press on to the fabric in about 30 seconds. This method is quick and efficient as the roller of paper design and the fabric roller can easily be replaced at the end of one run.

The advantage of transfer printing lies in the fact that the method eliminates the process of applying the colours of a design to the fabric in separate stages as in screen printing. Once the design has been transferred to the cloth the usual finishing processes are carried out to fix the dye.

Transfer printing is particularly of use in knitted fabrics constructed from triacetate, acrylic and polyester fibres. Garments already made up can also be individually printed in this manner and it has a particular reference to the manufacture of T shirts.

DYE MIXING FOR PRINTING

It is most important that during the printing of many yards of material there is a consistency of colour throughout the length. Formerly dyes were mixed in large vats or bins containing the gum, which forms a printing base; any additives necessary, and the dye, were added in a liquid measurement by an operator. Often this method was subject to human error. To avoid this situation a system of dye mixing has been developed which makes use of the computer.

The proportions of gum, additives and dye required are inscribed on a punched card and this is inserted into the computer. The vat to contain the printing ingredients is placed under a pipe and by pressing the appropriate button on the computer panel, the correct amount of gum is transferred to the vat from the pipe. The vat moves to the next pipe to receive the additives and moves on again to receive the correct amounts of dye from the appropriate pipes. The supply of gum, additives and dyes is stored in large containers above, and the contents of these travel down the pipes when required to the vat on a platform below.

Once all the ingredients are fed into the vat, the contents are transferred to a mixing machine to be thoroughly stirred. The mechanism, operating on a time switch, has rotary blades which descend into the vat and turn the mixture over, at the same time performing an up-and-down movement so that all the ingredients are mixed together within the depth and circumference of the vat.

CHAPTER 16

Printed and Dyed Styles

Direct Printing; Blotch; Mordant; Discharge Style; Resist or Reserve Styles; Batik; Tie and Dye; Printed Warp; Chintz; Cretonne

DIRECT PRINTING

In this method the colour is applied directly to the fabric by any printing method, e.g. block, screen or roller. The dye that has to be applied must be sufficiently thick not to spread beyond the area of colour required. To achieve this, a thickening agent is added to the dye, producing a very thick paste, and the agent employed depends on the type of dye to be used.

The dye, with which the cloth is printed, must be fixed in order to withstand the laundering, dry cleaning and general wear it will eventually receive. The fabric after printing and drying is steamed to allow the fibres to swell and the dye on the surface of the cloth to dissolve and penetrate. After the treatment with steam and acid baths, the thickening agent is removed by washing and rinsing. Direct colours, vat dyes, pigment dyes, chrome colours and basic dyes are used in direct printing.

BLOTCH

A term which refers to large areas of background colour that are printed by any known method. This can be seen on fabrics where the background colour surrounds a large motif. In preparing the blotch, the join is carefully staggered to avoid overprinting any of the colour. The staggering also removes the possibility of horizontal lines in the background, where one area of blotch joins another.

PRINTED AND DYED STYLES

MORDANT

The quality and fastness of most dyes can often be improved by a process known as mordanting, derived from the French word 'mordre' meaning to bite or fasten. The material is processed by steeping it in a warm solution of metallic salts, such as iron, aluminium, copper, tin or chromium, which combines with the dye and fastens it more securely to the fibre. Most fabrics with a guarantee against fading are subjected to mordanting.

DISCHARGE STYLE

The cloth is dyed to the colour required and printed with a reducing agent which removes the dye in the areas it is applied to. This leaves a white pattern on the original ground or fabric, and gives a background depth and evenness of colour that cannot be achieved by printing. A typical discharge cloth is a fabric with a coloured ground and white spots.

The reducing agent usually employed in discharge printing is sodium sulphoxylate formaldehyde, but caustic soda, tannic mordants, bromate of soda, and hydro-sulphate discharges can also be used. For a white discharge on wool and silk, it is sometimes preferred to use Redusol Z or Formosul C.W.

If a coloured discharge is required, dyestuffs that withstand the action of the reducing agent are included in the discharge paste. These pastes will then remove the dye from the coloured ground, and at the same time, deposit colour on the original ground; thus a fabric may be in two or more colours. The dyestuffs which are suitable for inclusion in the discharge paste are vat dyes for use on cellulose fibres and a selection of basic dyes for wool and silk.

RESIST OR RESERVE STYLES

In this method a substance is applied to the cloth that will prevent the fixation of any colouring afterwards employed.

In discharge printing the discharging agent is applied to the cloth after it has been dyed, but the opposite happens in resist printing as the reserving or resist agent is applied before dyeing takes place.

There are two classes of resist and in one method fats, resins, china clay, zinc oxide, sulphates of lead or barium are applied. The other type of resist employs chemical compounds, such as acids, alkalis and salts. The resisting agent used depends on the class of dye to be employed. In the china clay resist, the clay is used with acids and prevents the paste from spreading beyond the limits required.

BATIK

The resist is obtained by applying wax to both sides, or only one side of the cloth. It has no commercial use, but is a most decorative craft method of ornamenting cloth.

The hot wax may be painted, drawn, or printed on to the cloth. Parts or the whole of the material are coated with a mixture of paraffin wax and beeswax, and allowed it to dry thoroughly; the wax may then be cracked. When the cloth is dyed, the dye penetrates the cracks and gives a veined effect. In drawn or printed designs, the hot wax applied to the cloth is allowed to dry and the material is carefully dyed in cold water, to avoid melting the wax. The dye penetrates the unwaxed areas, leaving the cloth under the wax in the original colour. To obtain a variety of colours, wax may be applied to the dyed areas and redyeing takes place.

In Javanese batiks the 'Tjap' and the 'Tjanting' are used to apply the wax. The Tjap is a block made from copper strips and formed into an iron lattice held in place by solder. The hot wax is poured into the metal block and printed on to the cloth.

The Tjanting is a small cup made from thin copper sheet with a spout at one end and fitted into a short bamboo rod. The diameter of the spout decreases slightly at the delivery

Sanderson & Sons Ltd.

Plate 12. Batik. Screen print in twelve colours on 100% cotton. Width of material 48″/50″. Repeat 35½″ length, and full width of material.

end. Several Tjanting tools may be used for one design, each with a spout of a different size in diameter. The hot wax is poured into the cup and the design is drawn on to the cloth with the spout. (Further information may be found in specialised books on the subject.)

The wax is removed after the completion of the design by boiling water, and the cloth is washed, rinsed and dried. Batik designs may be adapted and transferred to screen and roller printing.

TIE AND DYE

In this form of resist the cloth, warp, or hanks of yarn are tied very tightly in parts with string and then dyed. The dye cannot penetrate beneath the string. When one colour dyeing takes place, the parts that have been tied remain in the original ground colour and the removal of the strings reveals a two-colour design. The cloth, warp or hank may be tied and dyed several times, and after the last dyeing the strings are removed to give a design with several colours.

For warp tie and dye, the warp is made in the usual manner and stretched to its full length. It is marked off in 6 mm, 12 mm or 25 mm sections, at both ends and in the middle. The areas to be left in the original colour are then tightly tied with string and the warp is dyed. Further tying and dyeing may take place.

After the last dye has been applied, the warp is dried and raddled (spaced out to the required width) and a number of ties are removed. Half the length of the warp is wound on to the loom and care is taken in aligning all the ties in the correct position widthways and lengthways. The remaining ties are removed and the remainder of the warp is wound on to the roller, to be entered through the shafts and woven in the normal manner. This method of tie and dye is purely a craft way to ornament the cloth.

PRINTED WARP

This method is referred to as chine or chene. The warp is

carefully run on to a warp beam by the usual method and a small amount of oleine oil may be applied to assist in the printing. In the printing process, the warp is passed in the form of a flat sheet to a beam, and as it travels round, the design is printed on to it. The printing machine is the same as for cloth printing. The warp is then transferred to the loom and is usually woven in a simple weave. The warp threads do not keep in the exact position so that the colours tend to merge and the edges become indefinite.

The unprinted weft may be all white or of different colours. The fabrics include silk, worsted, cotton and linen. Printed imitations are made of this technique, but a printed warp is easily recognisable by the removal of a few threads, which will show the warp printed and the weft plain.

CHINTZ

The term chintz refers to a fine, closely woven, cotton cloth usually printed with an elaborately coloured design. The surface of the cloth is highly glazed and it is used for furnishing purposes. It cannot be stitched very closely or the fabric will easily split along the line of stitches.

CRETONNE

The cloth is woven with a fine cotton warp and a thick, soft weft which is frequently condensor, spun from waste cotton. The weaves employed are plain, twill, oatmeal crepe or small fancy weaves and the cloth is usually lustreless. Cretonne is printed in various styles of design.

FLOCK PRINTING

In the electrostatic method of flocking, the cloth is printed with an adhesive and passes on a moving belt through an electric charge. The flock, made from cotton, rayon or synthetics is filtered from the flock hopper on to the cloth and is attracted to the adhesive in perpendicular form. After a baking process, the surplus fibres are removed.

CHAPTER 17

Preparation and Printing of Fabrics

Cotton; Linen; Wool; Silk; Viscose Rayon; Acetate Rayon; Nylon; Terylene

COTTON

Before the printing operation, the cloth must be entirely free of foreign matter and natural impurities, which if allowed to remain may detract from the purity and lustre of the finished cloth. The impurities consist of wax, fat and size used to assist in weaving the cloth, and the dust, dirt and oil from the loom.

A number of lengths of cotton are sewn together, and subjected to singeing and shearing to remove the surface fluff or fibres. After singeing the pieces are scoured and bleached, in some cases the scouring is a part of the bleaching process. The rope of fabrics is packed into a kier, a boiler which is hermetically closed, and the air is withdrawn. Caustic soda is pumped into the vessel from below and the liquor heated whilst circulation takes place. The process takes from 5 to 10 hours, and after cooling and rinsing, the fabrics are removed and washed; this may be preceded by scouring with acid. Kier boiling removes all the impurities and the fabric can usually be printed without further bleaching.

If the fabric requires further bleaching, hypochlorite acid is applied to obtain a full white. Material that has not had a kier treatment may be bleached with hydrogen peroxide, which is made alkaline with the addition of caustic soda, or sodium peroxide. A method of combining hypochlorite bleach with a peroxide bleach can be used, and this process may be after a kier boil or be carried out without one. The fabrics are washed, thoroughly rinsed and dried before printing.

Cotton fabrics are printed with direct, vat, chrome and basic dyestuffs as these give good light and wash fastness. Cotton is printed by any of the three methods given. It may also be printed with caustic soda and this causes the fibres to shrink in those parts where the acid is applied. By printing with agents that destroy the fibres a pattern of holes is made in the cloth.

LINEN

Linen follows lines of cloth preparation similar to those prior to printing cotton. The impurities in the fibre are more difficult to remove as a severe kier boiling would cause the fibres to deteriorate. A hypochlorite bleach is given and the fabric may have two kier boilings, one before the application of the bleach and one after bleaching.

The dyes used for cotton are suitable for linen and the printing methods are also the same.

WOOL

Wool has a greater affinity for dyestuffs than cotton, and when fastness is more important than brightness the faster shades of mordant colours are used. In discharge printing on wool, the cloth is dyed with direct colours and discharged with stannous chloride or hydrosulphite.

The woollen cloth is prepared for printing by scouring in alkaline baths, washing, and rinsing followed by passing it alternately through baths of bisulphite of soda and hydrochloric or sulphuric acid. The cloth is put through a solution of stannate of soda or chloride of tin and a weak solution of bleaching powder. Washing and rinsing take place and the cloth is dried ready for printing. In some instances, the tin preparation is omitted, and the material is chlorinated and bleached with bisulphite of soda, bleaching powder and sulphuric acid.

The printing paste used for woollen materials is always

thickened with substances which will wash out of the cloth readily in warm water. The thickening agents are generally gum senegal, gum tragacanth and British gums. The dyes are either acid or basic, combined with thickening and various acids, e.g. acetic, formic and oxalic.

Wool can be printed either by block, screen or roller. In the roller printing of wool, a design with heavy areas of colour and blotches must be deeply engraved, and the cylinder requires more padding than that needed for cotton printing. The woollen material is steamed, washed and rinsed after printing. The material is hydro-extracted and finally dried on a stentering machine.

Woollen and cotton mixture cloths can be printed with acid colours, which have no affinity for cotton and only fix on the wool. Or they may be printed with two dyes, one with an affinity for cotton and one with an affinity for wool. The preparation of this mixture cloth is the same as for an all-wool fabric prior to printing. In a fabric with a combination of wool and acetate rayon the cloth is printed with direct or acid dye for the wool and dispersed dyes for the rayon.

SILK

The printing of silk is the same as for the printing of woollen fabrics. The preparation prior to printing is different from wool, as the silk must be thoroughly cleansed to remove all traces of gum, and this is achieved by washing the cloth and afterwards mordanting it with tin. If the silk has a yellow tinge, it is usually bleached with hydrogen peroxide before the mordant is applied.

The dyes suitable for printing silk are basic, acid, direct and mordant colours. Block, screen and roller printing can be employed for the printing of silk, but care is taken with the latter method as silk is a relatively unabsorbent material and the colours can smear or streak in the machine. The thickening agents employed are those which wash out easily after printing and these are gum tragacanth, gum senegal, gum Arabic and British gum. For darker shades, gum tragacanth gives faster colours to washing and leaves the cloth soft and pliable, gum senegal gives good pale to medium shades and British gum is more suitable for very fine silk fabrics.

VISCOSE RAYON

The material is de-sized and washed before printing takes place. Viscose rayon has an affinity for all types of dyestuffs applied to cotton and the thickening agents used are the same as for silk. Direct cotton colours are principally used for the dyed ground, and hydrosulphite as the agent in discharge printing.

After the printing operation the fabric is steamed and care is taken to avoid any strain on the material, which would cause it to fray or break. The washing of viscose rayon is usually carried out by spraying the fabric as it travels along in open width through the washing machine; this avoids abrasion to the delicate structure. The cloth is rinsed and dried by cylinders or on a stenter frame.

ACETATE RAYON

In acetate rayon printing the cloth is subjected to a thorough scouring with a slightly alkaline soap to remove all the impurities before printing commences. Pure acetate rayon fabrics seldom need bleaching, but occasionally a mild bleach is applied to improve them. Mixtures of cotton and rayon require a bleach as it is difficult to obtain a pure permanent white on the cotton without bleaching. Scouring and bleaching can be carried out as one process, using hydrogen peroxide as the bleach.

Acetate rayon has an affinity for basic and indigosol dyestuffs, but none for the other groups of dyes, and the dyes developed for it are dispersed dyes. In fabrics composed of mixtures of acetate rayon with cotton, silk, viscose rayon or wool, the acetate dye is combined with the dye for the other fibre. These are evenly matched and the fabric is printed with them together. The acetate dyestuffs give a good fastness to light and washing when they are printed on all acetate material. The fabric can be treated to give it an affinity for other dyestuffs, such as direct and vat dyes, or by adding chemicals to change the fixing of the colours after printing.

The thickening agent used with dispersed dyestuffs has to have a low water content, as the acetate rayon does not absorb water readily. The low water content allows sharp outlines of pattern to be printed, which are not possible if the dye mixture is too wet.

By printing a design on the rayon with a delustring agent, a matt or imitation damask may be formed in contrast to the sheen of the background. All acetate fabrics are steamed to fix the dyes; careful washing, rinsing and drying follow.

NYLON

The nylon fabric is scoured to remove the size and oil acquired in the weaving process. The scouring bath contains soap or sulphated fatty alcohol or a mixture of these two and care is taken not to stretch the fabric warpways. Filament nylon must be heat set prior to printing in order to give stability to the fabric.

As nylon absorbs very little water, the thickening agent included in the dye mixture has a very low water content. In machine printing the rollers have a very shallow engraving to avoid any movement of dye on the cloth before it dries. The dyes used for nylon are dispersed, selected acids and a few direct dyes.

TERYLENE

Terylene is similar in preparation to nylon and has a low water content thickening agent in the dyepaste. Dispersed dyes are used for Terylene and the cloth after printing is steamed under pressure to obtain heavy shades, with an ordinary steaming for light shades.

CHAPTER 18

Preparation and Dyeing of Fabrics

Cotton; Linen; Jute; Hemp; Sisal; Ramie; Wool; Silk; Viscose Rayon; Acetate Rayon; Nylon; Terylene; Crimplene; Orlon; Acrilan; Ardil; Piece Dyed Cloth; Cross-Dyed Cloth

A dye is a substance which will colour a fibre, and adhere to it sufficiently, so that it is not removable by rubbing and washing. It must withstand the normal conditions of light, dampness, heat and air pollution. In dyeing a fabric as opposed to printing it, the former gives a complete application of colour throughout the length and width, and the latter is colour applied in pattern form.

For fabric dyeing the dye must be water soluble, or must be made soluble with the aid of chemicals. The dye solution is termed a dyebath or dye vat and this is made up from a percentage of dyestuff, water and acids. The amounts of these substances is calculated on the shade to be dyed and the weight of the material. For instance a 1 per cent dyeing will give a shade produced by 100 kg of cloth with 1 kg of dye.

To achieve good, even dyeing, it is better to exhaust the dyebath; this means that the material has absorbed all the dye, leaving the water clear. For some materials, exhaustion does not take place, and the dye remains in the dyebath after equilibrium, the point where the material is unable to absorb any more dye.

Each dyestuff varies individually and its behaviour depends on the type of fibre, the conditions of dyeing and other factors. As some dyes do not have an affinity for certain fibres, the affinity can be created by applying a mordant, which creates a combination between the dye and the fibre.

The dyeing of fabrics takes place in three stages. The first is the attachment of the dye to the surface of the fibre, the second is the penetration of the dye into the fibre, and the last is the fixation of the dye. In order to achieve these three phases, the fibres must swell to allow the dye to be absorbed, and this is carried out by the water in the dyebath and by the raising of the temperature. The dye, penetrating the fibres, is then fixed by the addition of salts and acids.

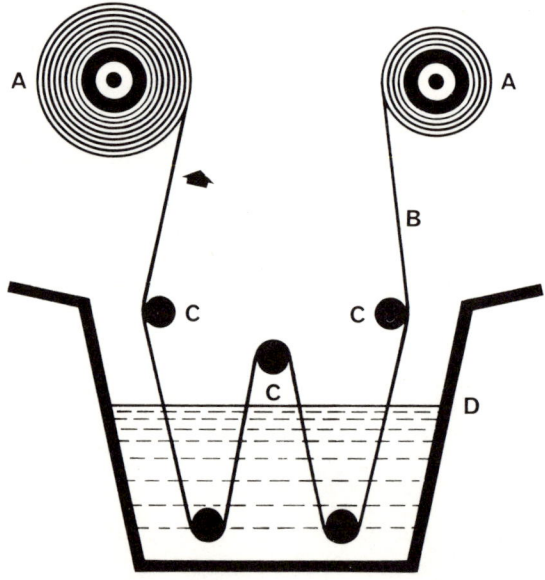

Diagram 48. Jigger Dyeing Machine. A = cloth roller B = cloth C = guide roller D = dye vat.

The dyeing of fabrics is referred to as dyed piece goods and may be carried out in the jigger, padder or winchbeck. The ordinary jigger consists of the dye vat, which is fitted with guide rollers. One end of the fabric is wound on to a cloth roller and the other end is attached to another cloth roller. The fabric travels through the dye liquor and winds on to one roller and is then reversed to wind on to the other roller. This process is continued until dyeing is completed and the cloth is rinsed by removing all the dye liquor and filling the vat with clean water. The cloth is lifted from the jigger on a batching up roller. In this type of dyeing there is considerable tension on the fabric and for delicate fabrics a tensionless jig is employed to avoid all strain.

The padder allows the fabrics to pass through guide rollers in open width, while another set of rollers squeezes the cloth. The compression of the rollers effects dye penetration and the cloth continues to travel through the dye liquor and compressing rollers until dyeing is concluded. The cloth then passes through a developing bath.

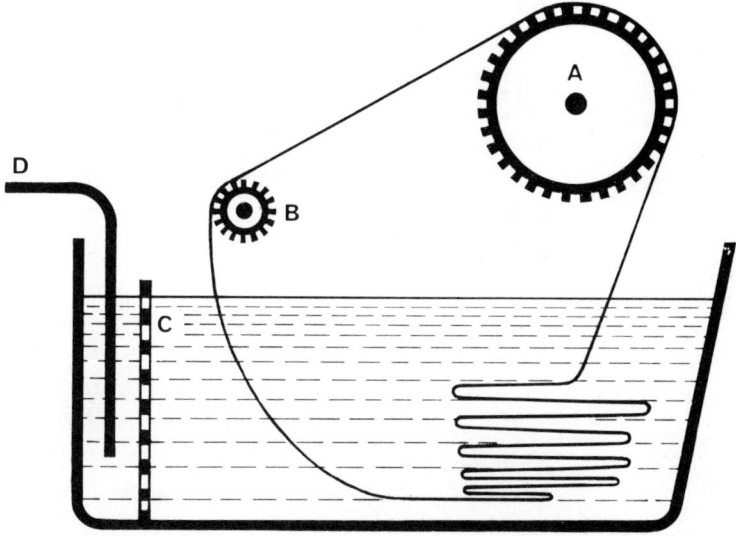

Diagram 49. Winchbeck Dyeing Machine. A = latticed roller B = guide roller C = perforated plate D = steam, water, chemical supply.

In the winchbeck method, the fabrics are sewn together to form a rope. A vat (the beck) contains the dye liquor and the winch, an elliptical or round latticed roller (A), pulls the fabric over a guide roller (B). The fabric drops in concertina form from the winch and slowly travels to the front of the beck, and is pulled up and over again. This process is continued until dyeing is fully effected. The beck has a perforated plate (C) that partitions off a space at the front, through which steam, water and chemicals are supplied (D).

With some fabrics it is possible to pad them with a suspension of dye, reducing agent and alkali, then develop the colour by passing the cloth through a steaming chamber.

COTTON

The cloth is thoroughly cleansed of all impurities, which, if allowed to remain, would result in patchiness on the finished article. The material is scoured and bleached, or both processes may be carried out as one operation. A kier boil is sometimes sufficient without the process of bleaching, but after either operation the material is thoroughly washed and rinsed. The dyes used in the dyeing of cotton fabrics are direct, vat, chrome, basic and sulphur colours with the addition of Glauber's or common salt.

LINEN

Linen materials are mainly used in their natural colour with only about 10 per cent being dyed. The dyeing of linen fabrics is carried out on similar lines to the dyeing of cotton. With 100 per cent linen fabrics, the dyes used are vat, solubilised vat colours and azoic; sulphur colours are used to a lesser extent. When the material consists of mixtures of linen and other fibres, the dyes are chosen which will have an affinity for both fibres.

The material is fully cleansed before dyeing and either half or fully bleached. As linen is a hard fibre it is more difficult to dye than cotton. A non-crease treatment is given to the fabric prior to, or after, dyeing with a solution of cold strong caustic soda.

PREPARATION AND DYEING OF FABRICS

JUTE

Fabrics woven from jute, mainly hessian, are dyed in a similar manner to cotton, as all dyestuffs suitable for the latter may be used on jute. The cloth is thoroughly cleansed, and if pale and bright shades are required it is bleached to a light cream with chloride of lime, or a combination of non-ionic detergent and sodium perborate. The more costly dyes such as vat are seldom used for dyeing jute fabrics. Sulphur colours are used for giving a good fastness to washing, water and light, but for carpet jute, the direct dyes are employed as they penetrate well and give a good fastness to rubbing and light.

HEMP

Hemp is not often dyed, but where it is necessary the dyes chiefly used are direct colours. For very bright shades the fabric is first mordanted with tannin and antimony and dyed with basic dyestuffs.

SISAL

Sisal has a good affinity for direct and acid dyestuffs and is usually dyed for the manufacture of matting.

RAMIE

As ramie is very similar to linen it may be dyed with those dyestuffs for which linen has an affinity, such as direct, basic, sulphur, azoic and vat. The fastness to light and washing for ramie dyed cloth is the same as for cotton or linen.

WOOL

The cloth is prepared by the processes used for wool printing. It is usually preferred to dye woollen cloth in a tensionless manner, and the dye liquor is luke-warm before the cloth is entered. As wool has a good affinity for dyes, especially acid and basic, it is an easy fabric to dye. The dyebath has a percentage of acetic or sulphuric acid added to it. A woollen fabric may be mordanted prior to dyeing, but this depends on the type of dye to be used.

When dyeing is completed, the cloth may only require rinsing or a mild soap wash. The material has the surplus water removed and is finally dried under tension on a stentering frame.

Materials which are woven with a mixture of yarns, e.g. wool and cotton, may be dyed with two dyes, one with an affinity for cotton and one with an affinity for wool, or dyed for wool leaving the cotton unaffected, and vice versa.

SILK

The cloth is prepared as for printing, with the removal of all impurities, and is washed and bleached. The dyes used for silk cloth dyeing are basic, acid and direct. If the silk has been boiled to remove the gum, the boiled-off liquor may be used for dyeing, by adding basic dyestuff and acetic acid. For acid dyes the dyeing process may be in two forms, acid dyeing by using boiled-off liquor with a percentage of sulphuric acid, or neutral dyeing in which clean water is used with the addition of Glauber's salts. Acid dyestuffs give a good fastness to rubbing and light, but are not so fast to washing. Basic dyestuffs may be used with tannin to assist the fixation of colour, and direct dyes are faster to washing, but less brilliant.

Silk is usually dyed without any tension as it is a delicate fabric and needs careful handling. After dyeing it may only need rinsing, or sometimes a mild soap wash and rinse followed by drying.

VISCOSE RAYON

The cloth has all impurities removed as for printing. Viscose rayon is of a fragile nature, especially when it is wet and heavy, and strain on it must be avoided. It is therefore dyed without tension. The fabric after dyeing is soft, rich in colour and has a lustre.

Viscose rayon material which has been woven in colour

may have marks or lines across the width. This is due to the yarn having a different dyeing affinity during the yarn dyeing process, and may also be caused by varying tension during weaving. The cloth can be dyed to level out the barring effect or variations in colour.

The dyes employed in viscose rayon dyeing are the same as those used for cotton dyeing, with the addition of Glauber's or common salt. Material which has been woven with plain cotton and viscose can be difficult to dye. Both fibres possess an affinity for direct colours, but they do not possess it equally. For dark colours the viscose absorbs more dye than the cotton and this is counteracted by dyeing at a low temperature, using as little Glauber's salts as is possible. With a mercerised cotton and viscose cloth, the effect is reversed and the cotton dyes darker than the viscose. This is hardly noticeable in dark shades, but in paler colours an excessive amount of salts is required to prevent unevenness.

Viscose rayon may also be dyed with dyes especially developed for it. The fastness of many dyes used for viscose rayon may be improved with after treatments on the cloth.

ACETATE RAYON

Prior to dyeing, the cloth is thoroughly cleansed in order to obtain an evenness of colour. The dyestuffs employed in acetate dyeing are mainly dispersed dyes and a few others.

The temperature of dyeing varies according to the type of fabric, as at a high temperature the lustre of the cloth is impaired. In dull or matt fabrics, the cloth may be dyed at boiling point. Delicate fabrics are dyed with care and must be free from all tension. On completion of dyeing, acetate fabrics are washed thoroughly to remove surplus dye, finally rinsed and drying takes place over steam heated cylinders.

Acetate can be pre-treated with dilute caustic soda and this gives it an affinity for direct dyes, but care is taken in dyeing as the soda can easily damage the cloth.

NYLON

The fabrics of filament nylon are heat set before dyeing, to provide stability. Nylon can be dyed with acetate dyes, acid, chrome and a few selected direct dyes.

TERYLENE

Terylene is set before being dyed at a high temperature to avoid any stretching or misplacing of the warp and weft. For the dyeing of Terylene the dyes used are dispersed dyes, and some azoic. It is difficult to produce heavy shades and this is achieved by high pressure dyeing.

CRIMPLENE

Similar in method to Terylene dyeing, and dispersed dyes give very bright shades and heavy shades. The fabrics have good fastness and level dyeing.

ORLON

Orlon is difficult to dye and pastel shades are produced with certain acetate and basic colours by dyeing the fabrics in the usual manner.

ACRILAN

The dyes used for wool may be used for dyeing Acrilan, and these are the acid and chrome range. Acrilan also has an affinity for acetate, basic, azoic and vat dyes.

ARDIL

As Ardil is very similar to wool it can be dyed with the same dyes in the same manner. Ardil absorbs and allows a deeper penetration of dye at a low temperature more quickly than wool. The dyebath is set with the required amount of dyestuff, Glauber's salt crystals and Acetic acid (30 per cent) and dyeing is carried out as for wool. The cloth is washed, rinsed and dried.

Chrome dyestuffs may be used for dyeing Ardil and the cloth may be mordanted with chromium salts prior to dyeing, or the salts may be added to the dyebath. In the mixing or blending of Ardil with cotton, acid and direct dyestuffs are used, either by blending the dyes and adding them to the dyebath, or by dyeing first for one yarn and then the other.

PIECE DYED CLOTH

This term refers to all cloth which is woven in its natural or grey condition and then dyed. This method is the most economical method of dyeing fabrics. A high proportion of cloth may be woven at one time, and by dividing it up into lengths of a specific number of yards, each piece may be dyed a different colour. This method disposes of the need to make many different coloured warps and enter each one on the loom. With piece dyed cloth it is possible to produce the same cloth in a variety of colourways.

CROSS DYED CLOTH

A term referring to a fabric when it has been woven with two different yarns, i.e. cotton and wool. One yarn may be used for the warp and the other for the weft, or it may be a blended yarn. The cloth can be dyed with a dyestuff for which only one yarn has an affinity, leaving the other unaffected. In some instances the cloth may be dyed first for one and then for the other, using two different dyes for which each yarn has an affinity, or two dyes may be blended together and both fibres dyed at the same time.

CHAPTER 19

Dyes

Classification of Dyes: Direct Dyes; Acid Dyes; Vat Dyes; Basic Dyes; Mordant Dyes; Sulphur Dyes; Pre-Metallised Dyes; Azoic Dyes; Mineral and Pigment Dyes; Oxidation Dyes; Dispersed Dyes

All primitive methods of dyeing fabrics and garments were based on natural colouring substances obtained from plants and vegetables. The early Egyptians used vegetable dyes, combined with mordants, and produced a yellow with safflower and iron pigments. A purple dye, used in ancient Greece for Royalty, was produced from sea snails, and the Britons at the time of the Roman invasion were dyeing wool and flax with colour extracted from the leaves of the woad plant. Very few vegetable dyes are used now as it is a more lengthy process than the dyeing of fibres with synthetic dyes.

The first synthetic dyes were made from coal tar products and were developed by a young British chemist, William Henry Perkin. The introduction of these coal tar dyes allowed fibres to be dyed in batches to exactly the same shade, which was more or less impossible with vegetable dyes.

The first coal tar dyes were used to dye only the natural fibres such as cotton, silk, wool and linen. It was found that silk and wool required a different method to the dyeing of cotton and linen. From these early experiments various dyes were developed to suit different natural fibres, and with the introduction of man-made fibres the research and invention of further dyes had to take place to suit the new fibres. Man-made fibres have also affected the equipment required for dyeing, as some of these need to be dyed at a higher temperature than the natural fibres. Dyeing apparatus has been made much more efficient, with the dyeing operation automatically controlled. There are a large number of machines for different types of yarns and fabrics.

Textile auxiliaries have also greatly aided the textile industry, as they assist in the fastness of the dyed article to washing and light. They give a deeper penetration and absorption of dye and assist the dyeing operation in many other ways.

The range of dye colours is unlimited, with new colours and methods being produced all the time, especially where man-made fibres are concerned.

CLASSIFICATION OF DYES

DIRECT DYES

A dye employed in the dyeing or printing of cotton, linen, and other cellulose fibres, some synthetic and acetate. They have excellent fastness to washing and light and are easily applied. Direct dyes are more readily absorbed by the addition of Glauber's salt or common salt to the dye liquor. They give level shades, and can be made more resistant to the removal of colour by washing, with an after treatment of formaldehyde or a solution of copper or chromium salts. Direct dyes can be used for wool under certain conditions.

ACID DYES

These dyes have a particular application in an acid bath to wool. The dyebath is made up containing the dye, and Glauber's salt with acetic acid or sulphuric acid added in stages. Acid dyes may also be used for nylon, Orlon, Terylene and Vinyon, and produce bright fast colours. Silk can be dyed with acid dyes by a neutral or acid dyeing method.

VAT DYES

A class of dye that gives shades with the highest degree of

fastness to light, washing and bleaching. In the dyeing of cellulose fabrics or yarn with vat dyes, the dyebath is prepared with the insoluble vat dye in fine powder form and a percentage of caustic soda and sodium hydrosulphite. As the dye dissolves the liquor changes colour (a blue dye can turn yellow) and when the material has absorbed enough dye, it is rinsed and treated with an oxidising liquor such as a solution of hydrogen peroxide, sodium bichromate or sodium perborate. This solution oxidises the material and gives it its true colour. In printing with vat dyes sodium formaldehyde sulphoxylate is used as a reducing agent.

Vat dyes have more affinity for cellulose fibres, but because of their excellent fastness, special methods have been evolved to allow their use in the dyeing of protein fibres.

BASIC DYES

Basic dyes are freely absorbed by protein fibres, but have a small affinity for cellulose fibres. The dyes have the power of forming salts with acids and are dyed with a small amount of acid. In the dyeing of cotton, viscose rayon and other cellulose fibres, it is necessary to mordant them with antimony tannate, Katanol ON or Taninol BM to give an affinity.

Basic dyes are brilliant in colour, but poor to light and washing fastness. Some improved types of basic dyes are used for Orlon and give light and wash fast colours.

MORDANT DYES

These dyes as the name suggests are used for dyeing fibres in conjunction with a mordant. In the 'metachrome' method of dyeing, the dye and mordant are applied simultaneously; this is mainly based on the use of chromium salts. A metallic mordant may be given before the dye is applied, or as an after process.

Wool is frequently dyed with mordant dyes when a superior colour fastness is required than can be obtained with acid dyes. The colour of mordant dyes depends on the type of mordant used, as each metallic salt will produce a different colour with the same dye powder.

SULPHUR DYES

In general sulphur dyes give rather dull shades, but they are inexpensive, easy to apply and have a good fastness. Sulphur dyes are somewhat insoluble in water and are heated in a solution of sodium sulphide, which gives a strong affinity for cellulose fibres.

In the dyeing of protein fibres the dye liquor contains sodium sulphide with an addition of sodium bicarbonate or some other substance which will neutralise the bath. These are added before the fabric is entered.

PRE-METALLISED DYES

Pre-metallised dyes contain chromium or copper and can be used for wool dyeing with a higher concentration of sulphuric acid added to the dyebath. In the dyeing of cotton and wool with pre-metallised dyes, the colours can be improved by an after treatment of a metallic salt.

These dyes are useful for the dyeing of man-made fibres, such as nylon and Orlon, but require special conditions of dyeing for deep shades.

AZOIC DYES

The greatest use of these dyes is in the dyeing of cotton, viscose and acetate rayon, synthetic fibres, and for silk, when a maximum amount of colour fastness is required.

Several components were developed for combining with the dye to form colouring matter within the fibres. These compounds, termed naphthols, can be applied to the fibres to produce a wide range of colours. The Naphthol AS was one of the first and other compounds have since been developed including a range by I.C.I. Ltd., called Brenthols. The Naphthol AS has a strong affinity for cellulose fibres and the dye can be applied in the same manner as a direct cotton dye.

The addition of Fast Diazo Salts dissolved in water, combined with the Naphthol AS compound and brought together, forms the insoluble dye within the fibre. In the printing of azoic colours the Naphthol AS compound and the diazotised base are steamed or otherwise treated so that the two dye components come together and form the dye.

MINERAL AND PIGMENT DYES

These dyes were originally not much used for the dyeing of textiles as they were not water-soluble and were applied to fabrics by printing with a suitable binder.

New binders based on synthetic resins are now used for dyeing materials and combine with a dispersion of coloured pigment and become water-soluble at an intermediate stage. The material is dried and heated; this process of baking the fabric causes the pigment to be bound to the fibre surface.

This method gives very bright colours, which are fast to light and washing. Glass fibres are dyed by this method, as they do not have an affinity for ordinary dyes.

OXIDATION DYES

With this method the material is impregnated with a substance and treated with oxidising agents. The substance, which is usually non-coloured, is converted into a pigment colour within the fibres and is then washed to remove unused chemicals.

Aniline Black is the dye most employed today for the dyeing of cotton, viscose and acetate rayon fabrics. It produces, very cheaply, a good black with an excellent fastness. Cotton fabrics are first impregnated with a liquor made up of aniline, hydrochloric acid, sodium chlorate and copper chloride or sodium ferrocyanide. After drying the fabric is steamed to bring about the action of dyeing and the cloth emerges black. To prevent a green tinge the fabric is run through a weak solution of sodium bichromate.

Aniline black is used for dress fabrics, umbrellas and scholastic gowns.

DISPERSED DYES

A dye developed for dyeing acetate, and used for nylon, Orlon, Vinyon N, Terylene, Crimplene and other synthetics. Dispersed dyes consist of pastes or suspensions of dye powder, which are dispersed in water with the aid of agents.

In the printing of fabrics, the liquid disperse dyes are stirred into the prepared stock paste. Once the liquid dye has dried up it cannot be redispersed, and to avoid this it must be stored in an even temperature.

These colours give a good fastness to washing and light, and produce clear, bright colours. With the addition of certain agents the steaming time can be cut down and the colour is better with pressure steaming on man-made fibres.

CHAPTER 20

The Care of Fabrics and Garments

Washing; Starching; Drying; Ironing. Natural Fibres: Cotton; Linen; Wool; Cashmere; Silk. Man-made Fibres: Viscose Rayon; Durafil; Sarille; Vincel; Acetate Rayon; Dicel; Tricel; Nylon; Bri-Nylon; Celon; Enkalon; Terylene; Crimplene; Fibrolane; Courtelle; Acrilan; Teklan; Glass Fibre. Laundry; Dry Cleaning. Removal of Stains: Mildew; Grass Stains; Scorch; Blood Stains; Fruit Stains; Coffee or Tea; Ink; Paint or Varnish; Grease or Oil; Wine; Rust (Iron Mould). Dry Cleaning of Stains

All garments should be washed or dry cleaned frequently, as over-soiling leads to fibre disintegration. The method by which curtains, clothing, knitwear and all types of furnishings are washed is an important factor in the life of a fabric. So many materials are ruined by the wrong type of washing.

The opening up of launderettes, although time saving and convenient for the housewife, has led to the incorrect washing of many furnishings and clothing. It is just not possible to wash all fabrics at the same temperature and with the same washing powders. Woollen knitwear should never be washed in very hot detergent water with other fabrics, as this damages all wool, causing it to shrink and felt, and leaving it harsh and unyielding.

Most manufacturers give washing instructions on a label inside the garments and it is better to follow this advice in order to prolong the life of the article. If the garment is not labelled or if there is any doubt about the fibre content or properties of the fabric, it is advisable to apply the following rules.

WASHING

Wash frequently, otherwise a harsh wash has to be applied to clean the fabric, and this can lead to damage to the fibre structure. The water should not be very hot and soap

THE CARE OF FABRICS AND GARMENTS

flakes or washing powders must be completely dissolved before the article is entered. Detergents and other powders should always be used according to the manufacturers' instructions. White articles should always be washed alone, as coloureds can have a slight loss of colour, which would tint the white clothes. Coloured fabrics and garments should be washed, rinsed and dried without interruption, as any soaking or prolonging of the washing process can cause loss of colour.

All fabrics should be rinsed several times to remove the soap suds completely, and gently squeezed to remove excess moisture, or if wringing is necessary a rubber wringer is preferable. Obvious man-made fabrics should be hung straight from the rinsing water without squeezing, as this can leave crease marks on the fabric after it is dry. For spin drying a short time is advisable as over-spinning may dry the fabric too much, making ironing difficult.

STARCHING

In domestic starching, the substances used penetrate into the spaces between the threads of the cloth, and impart a gloss on being ironed. Starching helps to prevent a garment from soiling quickly and the dirt can be removed easily from the starched fabrics without damaging the fibres.

Starches are made from rice, maize and wheat, and all natural starches are insoluble in cold water, and must have boiling water added to activate the solution. A chemical starch available is mixed with water according to the stiffness required in the fabric. This starching will last for several washes without having to be repeated. There is also a spray-on starch which is easy to apply before ironing.

DRYING

The garments or curtaining should be hung to dry as soon as possible, as damp garments rolled or folded up in a warm atmosphere can become mildewed if they contain natural fibres. For knitted garments excess moisture should be removed either by a short time in the spin dryer, or with gentle squeezing. This can be followed by laying the garment out on a towel in its correct length and width and the surplus moisture patted off with another towel on top. They can then be hung on a line or left to dry on the towel. Knitwear should never be hung very wet on a line as this will stretch it.

IRONING

To avoid imparting a sheen to fabrics, they should be ironed on the wrong side. A hot iron can damage man-made fibres, causing them to shrink or shrivel up. It is advisable to set the iron at a fairly low temperature and increase the heat if necessary. Some garments which may appear to be of one yarn only, can in fact be made from a blended one, and the iron should always be set for the fibre requiring the lower temperature. Heavy fabrics should be pressed under a damp cloth or with a steam iron. If garments are too dry, re-wetting is more satisfactory than sprinkling drops of water, as this can mark some fabrics.

NATURAL FIBRES

COTTON

White cotton materials will withstand a very hot wash or boiling, with a reliable soap powder or detergent. A mild bleach may be used, but this is quite unnecessary if the garments are well washed and rinsed. They should be ironed while still slightly damp, or if too dry can be sprinkled with water. The iron can be fairly hot and the cotton can be starched prior to ironing for a stiffer cloth.

Coloured cotton materials can be washed together, providing that the colours are fast; strong detergents should be avoided, and clothes must not be left soaking as patchiness or loss of colour may result. To prevent colours running or fading, the addition of common salt to the washing liquor will considerably lessen this fault.

THE CARE OF FABRICS AND GARMENTS

Cotton materials may be put through a wringer or spin dryer and hung up to dry. Drip dry cottons should be washed in soap powder or detergent with plenty of water. The fabrics should not be squeezed during washing and should be rinsed thoroughly without crumpling the fabric. Dresses and shirts should be hung on hangers and pulled into shape before being left to drip dry.

LINEN

Linen can be washed as cotton when it is made up into sheets, tablecloths, napkins, etc., but in its natural colour it should have a less vigorous wash. Dyed linen should be carefully washed with as little squeezing as possible. Linen can be ironed damp with a fairly hot iron.

WOOL

CASHMERE

A hand hot wash, with pure soap flakes or synthetic fluid specially prepared for woollens, is required. The soap flakes should be completely dissolved before the garment is entered. A gentle agitation is permissible while washing in two changes of water, and a thorough rinsing in warm water. A short time in the spin dryer is followed by hanging the garment on a line to dry, making sure that it has not been pulled out of shape. Knitwear can also be dried flat on a towel by patting out excess moisture on another towel. Woollens should not have any wringing.

Wool has a tendency to shrink and should not have very hot water applied, too much soap, or an excess of agitation in the washing and rinsing processes. All fabrics made from animal fibres should be washed alone. If knitwear requires pressing this should be carried out using a damp cloth or steam iron.

SILK

Silk should be treated gently and washed as wool, and

ironed while still damp with a warm iron. The garments or fabrics should be re-wet rather than sprinkled with water if they are too dry, as sprinkling can mark the silk.

MAN-MADE FIBRES

VISCOSE RAYON

DURAFIL

SARILLE

VINCEL

These fabrics are easy to wash in hand hot water, approximately 48°C or 118°F. They should be squeezed gently, when wet, as hard wringing could damage them. A reasonably hot iron may be used on the wrong side of the fabric.

ACETATE RAYON

DICEL

All acetate fabrics should be washed frequently in only warm water, as very hot or boiling water can be damaging. They should be rinsed thoroughly and squeezed gently, and ironed with a warm, not hot iron, on the wrong side.

TRICEL

Clothes made from Tricel have a good soiling resistance and dry quickly, needing little or no ironing. The fabrics require a hand washing in warm water for the minimum iron garments, without squeezing and wringing, and are hung up very wet to drip dry.

Blended or mixture yarns of Tricel and other fibres can be washed in a washing machine, providing the other fibre is suitable for this treatment. Wringing and spin drying should be avoided, but tumbler drying is satisfactory. When ironing is required, the garments should be ironed very damp or with a steam iron.

THE CARE OF FABRICS AND GARMENTS

NYLON
BRI-NYLON
CELON
ENKALON

These fabrics should be frequently washed in hand hot water, with soap flakes or synthetic detergent. Thorough rinsing must take place or the nylon will become discoloured. Nylon should be washed separately from other fabrics and white nylon should be washed alone and not with coloured nylon.

The fabrics can be drip dried or excess moisture blotted off before hanging to dry. Any ironing must be a low setting.

TERYLENE

A hot wash of 48°C, 118°F, either in machine or by hand is required for Terylene fabrics. Excessively soiled spots can be carefully rubbed with an extra soap or detergent application to that area.

The rinsing water can be shaken off the fabrics or they can be rolled in a towel before hanging up to assist in the drying. If ironing is required the iron should be set at low.

CRIMPLENE

Crimplene garments can be washed in hand-hot water either by washing machine or hand. It is advisable to wash Crimplene garments on their own and to turn them inside out before washing. Pleated garments should be washed by hand and given a drip dry. Ironing is not usually necessary, but if it is desired a damp cloth and low set iron may be used.

FIBROLANE

All fabrics with Fibrolane blended fibres should be washed as wool and ironed according to the requirements of the other fibre.

COURTELLE

ACRILAN

A warm water wash is required and the fabric should be thoroughly rinsed. Wringing is unsuitable and a gentle squeeze to remove excess moisture, followed by hanging the fabric to drip dry, is advisable. Some garments may need a light press with a cool iron.

TEKLAN

Garments made from Teklan are easy and safe to wash and require little if any ironing; a cool iron should be used if it is felt necessary. They dry quickly in the usual conditions of domestic laundering.

GLASS FIBRE

Wash the fabrics in a mild soap or detergent in the maximum amount of water at a low temperature. Glass fibre fabrics must not be bleached. The fabrics should be rinsed thoroughly and given a short time in the spin dryer and should be hung on a padded line to dry. They should not be ironed.

LAUNDRY

Most laundries subject cotton and linen goods to a mild bleach of sodium perborate or a slightly stronger one of sodium hypochlorite. Table linen is bleached to remove stains, which cannot be removed by the usual washing methods.

A launderer's process of washing wool is the same as for the domestic method. Some silk fabrics tend to shrink and it is often advisable to have these dry cleaned.

In sending any garments or curtains, covers, etc., to the laundry and where there is any doubt of the washing properties, it is advisable to label them, stating the fibres used. This will then avoid any errors in washing. Glass fibre fabrics should be washed as for wool and a domestic wash is usually more suitable than laundering.

DRY CLEANING

Curtains, covers, coats, suits, dresses, etc., are cleaned in automatic machines with highly refined cleaning fluids.

All articles are sorted according to the type of fabric, the colour, and weight. Evening and cocktail dresses, or articles made from delicate materials, such as silks, light woollens and man-made fibres, are given a special treatment. Those that are heavily soiled with dirt or oil are also cleaned separately, particularly overalls and dungarees.

The cleaning fluid is generally Perchlorethylene and the clothes are gently tumbled in a machine so that there is a constant flow of the liquid through them. The dirt is removed, rinsing in clean spirit follows and the clothes are dried by gentle tumbling in warm air.

The articles are steam finished and pressed in special appliances, each one selected according to the type of fabric. In some instances, hand ironing is used as this is the only method suitable for some articles.

Re-texturing is a dressing applied to fabrics to restore the appearance if signs of wear are present. Moth-proofing and flame proofing can also be given and will last until the next cleaning.

The re-dyeing of garments is carried out by dry cleaning firms although the choice of colours is rather limited. The re-dyeing of acrylic and other man-made fibres is usually impossible, except for pale shades. Most types of carpets can be dyed, but it is not possible to obscure an existing pattern.

If garments or furnishings do not have a label attached to show the type of fibre, it is usually advisable to label the articles if the fibre is known. If this is not possible it is wise to seek the advice of the dry cleaning firm before submitting any article for cleaning.

REMOVAL OF STAINS

All stains on fabrics should be treated gently, as hard rubbing, or the wrong application of cleaning fluid, can ruin the cloth.

MILDEW

The stains may be removed by several methods, and can be treated repeatedly with a solution of sodium hypochlorite and exposed to the air, then washed and rinsed. For newly formed stains a strong soap solution with exposure to the air is recommended. A solution of hydrogen peroxide or sodium hypochlorite can be used on non-coloured fabrics. With dyed fabrics ammonia and acetic acid in dilute form may be applied.

GRASS STAINS

Apply an extra application of soap to the affected part and the stain may be removed by rubbing. Obstinate stains may be removed by applying hot methylated spirits, or hot glycerine followed by methylated spirits. On white materials hydrogen peroxide with ammonia is effective.

SCORCH

An old-fashioned remedy of scraping the stain or burn with a silver coin with a milled edge is often successful, particularly on carpets. A vigorous brushing with a stiff brush will remove some scorch marks, but care has to be taken so that the surface of the fabric is not damaged. For more fragile fabrics the stain may sometimes be removed by washing with an extra application of soap to the scorched area. White fabrics can have a mild bleach applied to the scorch marks.

BLOOD STAINS

Soak in cold water, or apply cold water if the fabric cannot be soaked. Wash with soap and cold water, rinse many times in cold water and then give the fabric a normal wash, and allow it to dry. On white fabrics, hydrogen peroxide

THE CARE OF FABRICS AND GARMENTS   201

with ammonia may be applied. Stains on silk may be removed with strong borax water.

To loosen old dried stains the fabric can be soaked in a gallon of water to which has been added 2 tablespoons of ammonia. A strong salt water solution can also be successful by adding 2 cupfuls of salt to 1 gallon of water.

FRUIT STAINS

Remove with boiling water if the stains are new, and a solution of hydrogen peroxide with ammonia will remove stains on white fabrics. Soaking in hot water and washing followed by several rinses can be applied to coloured fabrics.

COFFEE OR TEA

These stains can be removed by applying boiling water, followed by washing and rinsing. Glycerine applied for old stains and completely removed by washing is usually successful. White fabrics may be bleached with hydrogen peroxide and ammonia with dilute acetic acid and rinsed in water.

INK

For ordinary writing ink a warm dilute solution of oxalic acid or potassium permanganate followed by hydrogen peroxide, is suitable for white fabrics. In coloured fabrics an application of petrol or turpentine is more usable. These substances should be removed with washing and rinsing.

BALL POINT PEN INK STAINS

Treat as a paint stain (below) before washing or cleaning and blot carefully with toluene and a light paint remover. If the garment is dry cleaned the stain must be pre-spotted and no stain should be visible after. If pre-spotting is not carried out prior to cleaning a fixation of the stain may take place. The stain becomes insoluble during the cleaning and tumbler drying due to oxidation taking place in the machine.

PAINT OR VARNISH

These can be removed with turpentine or petrol followed by a complete wash and rinse.

GREASE OR OIL

Place a clean cloth or blotting paper under the stain and dab gently with benzine, petrol, or a proprietory stain remover with a clean cloth. Wash and rinse thoroughly.

WINE

Wine stains should be easily removed with a normal wash, providing the stain is not allowed to dry in to the fabric.

RUST (IRON MOULD)

The stain should be moistened with lemon juice or a solution of salts of lemon or salts of sorrel and rinsed at once in water containing a little ammonia.

DRY CLEANING OF STAINS

If stains cannot be removed easily with any of the given methods, it is advisable to have the fabric dry cleaned. Any reliable cleaning firm will do its best to remove the stain without damaging the fabric. Staining liquids on man-made fibres can generally be removed if blotting paper or a soft cloth are applied at once to absorb the liquid. These fibres do not absorb moisture as quickly as natural fibres, and this makes staining easier to remove.

Petrol and many other cleaning substances are highly inflammable and should not be used near a naked flame—this includes lighted cigarettes. They also give off fumes and ideally it is best to apply them in the open air. If they are used indoors, the room should be well ventilated.

CHAPTER 21

Knitted Fabrics

Knitted Fabrics; Plain Knitting; Purl Stitch; Rib Stitch; Float Stitch; Tuck Stitch; Tuck Float Rib; Plated Stitch; Accordion Stitch; Jacquard Designs; The Latch Needle; The Bearded Needle; Fly Needle; Straight Bar Machine; Flat Bar Machines; Flat Bar Purl Machines; Circular Machines; Warp Knitting Machines; Gauge; Single Jersey; Double Jersey; Stretch Fabric; Tricot; Locknit; Interlock; Atlas; Milanese; Lace; Angel Laces; Ribbon Slot Lace; Curtain Nets; Marquisette; Tulle or Dress Net.

This type of fabric construction has become more important in the last few years. Knitted fabrics are now used for a variety of clothing and furnishings. For certain purposes a knitted construction is more suitable than a woven one and it will stretch and give greater elasticity than the woven fabric.

The commercially produced knitted fabrics are simply a mechanical extension of hand knitting. The principle of hand knitting and machine knitting is the same and this is the interlacing of a yarn in loops to form a fabric. With hand knitting the yarn is interlaced with the aid of two needles for a flat fabric and four needles for a tubular fabric. A row of loops are on one needle, the other needle is pushed through the first loop, the yarn is wound round between the two needle points and over one needle. The first loop is slipped over the yarn and cast off, leaving the new loop on the needle.

In machine knitting as in hand knitting the fabrics are produced as either a flat fabric or a tubular fabric. Each loop is on a small needle and moves up the needle when the yarn is fed in to form the new loop. The old loop is cast off as the new loop is pulled through. The interlacing of the yarn to form the loops can be made with a single yarn or thread where the loops are made horizontally, termed weft knitting, or from a set of threads placed side by side, known as warp knitting. These terms are derived from weaving, the horizontal threads referred to as the weft and the vertical threads known as the warp. A horizontal row of loops in the knitted fabric is known as a "course" and a row of loops lengthways is known as a "wale". The number of wales per inch depends on the knitting needles per inch employed and the count of yarn and the number of courses per inch is an indication of the stitch length. Finishing processes alter the number of courses and wales per square inch, and in some fabrics an overfeeding of yarn will produce a cloth which can be stretched afterwards to obtain a greater width. Selvedges are formed as in weaving on flat fabrics, and in weft knitting on the circular machines, some needles are taken out at a given point to indicate a cutting line. In designing knitted fabrics, various symbols are used to denote colours or stitches.

KNITTING STITCHES

PLAIN KNITTING

With this stitch a single jersey fabric with V shaped stitches on the front of the cloth and semicircular loops on the back is produced. The U shaped semicircles are the sinker loops and the inverted semicircles are referred to as needle loops. Plain knitting is the most widely used stitch and presents a smooth, even-faced fabric. A plain knit fabric normally has up to 40 per cent stretch in the width. If a stitch is broken that wale row becomes unravelled and a fault resembling a ladder occurs. Plain knitted fabric can easily be unravelled from either end.

PURL STITCH

For purl stitch the plain structure is reversed in each course and this produces a fabric which looks identical on both

KNITTED FABRICS

sides. The loops are knitted to the front and the back. The simplest purl structure is 1/1 and alternate courses are knitted in opposite directions. Thus courses 1 and 3 are knitted to the back and courses 2 and 4 to the front. Patterns or designs are produced by selecting the loops on each course for either front or back knitting. A 1/1 purl fabric has a lengthways stretch and can be unravelled from either end.

RIB STITCH

A stitch formed by two sets of needles arranged opposite each other, but fed by the same thread. If all the needles are used a 1/1 rib is knitted. The needles of the two sets are arranged so that some loops are knitted to the front and others to the back. For a 2/2 rib, two needles of one side alternate with two needles of the other side. There are many types of rib 3/1, 4/1, etc., the first number indicates that consecutive wales are knitting to the front and the second figure those wales knitted to the back. Broad rib fabrics have greater elasticity than a 1/1 rib structure. The 1/1 rib has nearly twice as much elasticity as plain knit and has equal loop tension on each side of the fabric. Rib structure fabrics can only be unravelled from the end knitted last.

FLOAT STITCH

A float or miss stitch is made by a needle retaining the loop it already has and failing to rise to take a new loop. This type of stitch is used in Fair Isle, Jacquard fabrics and other types of design in colour. The designs are carried out carefully to avoid long horizontal floating threads, as these would restrict elasticity in the fabric.

TUCK STITCH

A widely used fancy stitch which is made by needles rising to take new loops while retaining the old ones. The loops accumulate on the needles and are later discharged together. The number of consecutive tucks depends on the number of loops that the needle can hold, and the maximum number is usually five.

In double jersey fabric, tuck stitch is used as a figuring effect, and with single jersey the stitch makes the fabrics heavier. In self coloured fabrics it is used to give a lacy appearance to the cloth and the lacy look is formed with a small hole and interesting thread distortions.

TUCK FLOAT RIB

In this method of knitting, consecutive tuck stitches are made in which a knop effect is produced. One set of needles knit for 1/1 rib while another set of needles produce either tuck or miss for two or three courses followed by all needles knitting on the next course.

PLATED STITCH

A fabric composed of plain structure with an inlay yarn, which is positioned between two plated yarns. The inlay yarn floats three and tucks on the fourth stitch and the yarn knitted to the back of the fabric is called a binder thread.

ACCORDION STITCH

Produced from a combination of knitting, tucking and floating. In a straight accordion the floating threads are eliminated by tucking which is always made on the same needles. The needles making the tuck alternate with those that knit and float. For selected accordion the design on the face side of the fabric is made by selecting either to knit or float. The float threads are eliminated by the use of tuck stitch, but can appear wherever required according to the design. In alternate accordion the floating threads are eliminated by alternate needles tucking. The needles which tuck on odd courses, then miss at even courses.

KNITTED FABRICS

JACQUARD DESIGN

These are produced by a system of knit, miss and tuck. Two sets of needles are used, one series of needles form the face fabric and the floating threads are eliminated by another set of needles which are in action at each course. In the simplest type of Jacquard, one course of design requires two knitting courses. The needles are selected for face knitting; those that knit at the first course, miss at the second course and those that miss at the first course knit at the second. In this way one row of design is produced in two colours, and for three colours in each course three courses are required to produce one row of pattern. In double jersey a Birdseye backing gives a more even tension to both faces of the cloth and is produced by alternate backing needles which knit at alternate courses.

The designs for Jacquard fabrics are carried out in the same manner as designs for a woven Jacquard. A graph paper is used taking into account the height and width of repeat, depending on the type of machine used, and these factors are relevant to every type of cloth. Each small square on the graph paper represents a needle and each square is filled in with a symbol to denote the colour or stitch. A design could be shown by X in one square, ● in the next square and a blank square to indicate a design in three colours. A horizontal line of small squares indicates one knitting course.

TYPES OF NEEDLES

THE LATCH NEEDLE

This type of needle consists of a hook at the top, and the space between the hook and the stem of the needle has a free swinging latch. The latch opens and closes to hold a new loop on the hook while casting off or knocking over an old loop. Latch needles move individually in grooves and the grooves are referred to as tricks.

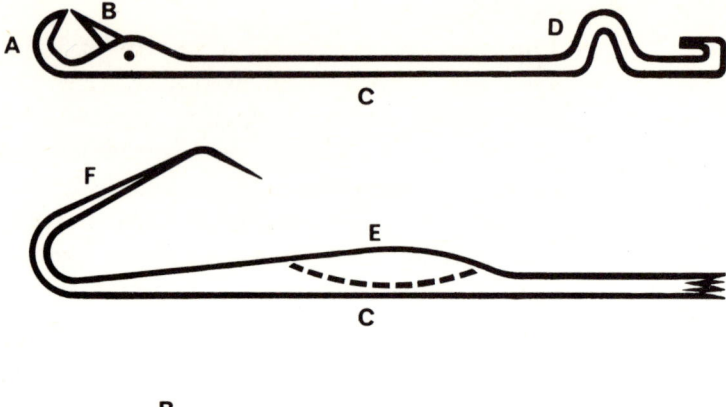

Diagram 50. Top to bottom: latch needle, bearded needle and fly needle. A = hook B = latch C = stem D = butt E = groove F = beard.

THE BEARDED NEEDLE

This needle is a hook with the top curved back towards the stem and ending in a point. The needle is a single unit and the closing and opening of the hook is effected by the point of the beard being pressed into a groove in the stem of the needle. This groove is called the eye of the needle.

FLY NEEDLE

A needle similar in construction to the latch needle. The closing and opening of the hook is achieved by a tongue moving up and down inside the hollow stem. This movement is carried out by a butt which interacts with the cams to open and close the tongue.

KNITTING MACHINES

A variety of machines are used in the knitting industry to produce either a flat piece of fabric or a tubular fabric. Every knitting machine has three basic units, the yarn feeder, the knitting bed and a take down mechanism. In

KNITTED FABRICS 209

every machine, one feeder represents one operation and there is one needle for every row of loops down the fabric. Machines produce either weft knitted cloth or warp knitted cloth.

STRAIGHT BAR MACHINE

The straight bar knitting machine produces flat fabric pieces which are knitted to the shape required. This system produces high quality fabric, and although slower in output, it is one of the leading methods used. It is employed in plain and rib knitting for outerwear and underwear.

In the straight bar knitting frame the bearded needle is used. The sinker and presser are most important, as the beard must be pressed into the groove to allow the old courses to slip over the newly laid thread. The bearded needles move together so that a complete course of loops is made at the same time. When the needles are positioned, the sinker forms the new loop and holds the fabric down wherever the needle rises. The presser bar provides a surface for the needles to press on to and grooves in which the sinkers slide.

The fabric is produced by the yarn carrier moving across the front of the sinkers with an additional distance in excess of the width of the fabric. As the yarn is fed to the needles all the parts of the machine are stationary and the needles are at their maximum height. The yarn carrier feeds in the yarn and alternate sinkers come forward and form the yarn round alternate needles, thus making one long sinker loop to every two needles. On the next movement, termed dividing, the remaining sinkers come forward and sink loops round all the needles. At this second stage all the needles move back slightly and begin moving down; this prevents strain on the fabric.

This process of feeding the yarn in and the formation of the loops is called the draw. The length of the stitch is decided during the draw. Sinkers always move into the

Nova Knit Ltd.   Plate 13.   Full double jersey 18 gauge knitting machine.

KNITTED FABRICS

same position, but the needle position can be adjusted so that the nearer the needles are to the presser bar, the greater the length of yarn on each stitch. When the needle has moved down and forward to the presser bar, a new loop is formed under the beard. As the beard is pressed, the loop is put on the needle beard and the sinkers move back to drop the new loop as the needle descends. The needle in descent comes below a knock-over unit and casts off the old loop. Simultaneously the sinker moves in to hold the loop down as the needles rise to begin a new course.

In a straight bar rib frame additional horizontal rib needles are used. The presser is formed so that the rib needle presses on the front underneath side. The frame needles move in to the press and the sinker moves back to control the new loops. When the beard has been pressed, the needle moves down with the beard still closed. The old frame loop is put on to the beard and as the frame needle goes down, the loops are drawn over the rib needles. The new rib loops are nearer to the presser than the previous loops formed. As the frame needles are in a low position, the rib needles move out slightly, bringing the new rib loops under the beards. Rib needles rise and press on the underneath side of the presser and the needles with the beards still closed move out so that the old rib loops are put on to the top of the beard on the rib needles. As the rib needles move out to knock over, a slide is moved in to assist in the casting off of the old loops. At the same time the frame needles move down slightly and in the second knock over the loops are cleared. The rib needles move towards the pressers, while the sinkers move out and hold the fabric down allowing the frame needles to rise and restart the process. This method is used for knitted outerwear as it is efficient and gives a minimum of fabric wastage.

FLAT BAR MACHINES

With this type of machine great scope is obtained for large

Nova Knit Ltd.
Plate 15. Patterning unit for double jersey machine.

Nova Knit Ltd.
Plate 14. Cylinder needle bed showing needle butts.

or small patterned fabrics. The machine is operated by the needles having individual movement and being positioned in grooves called tricks which are cut into a flat bar. The latch needles in the machine move vertically up and down and this is effected by the cams in the cam box, sliding along the needle bed. The yarn carried by the thread carrier moves along the length of the needle bed, and the needles embedded in the needle bed retain an open position until the new yarn is fed in. When the yarn has been placed in position, the needles move to their lowest point, drawing the new loops through the old loops which are then cast off.

For rib knitting, two sets of needles normally positioned at right angles to each other are required, and although latch needles are most common, bearded needle machines can be used. In a 1/1 rib the needles are synchronised alternately so that the needle in one needle bed is opposite to the tuck wall in the opposite bed. Other rib structures are formed by setting the needles in each bed; this may consist of empty tricks to give sufficient space for needles in the opposite bed. These settings are carefully worked out to give an evenly knitted fabric.

FLAT BAR PURL MACHINES

In purl knitted fabrics the general method is with double ended needles which can be transferred; this allows the loops to be knitted either to the right or left. Flat bar purl machines have horizontal needle beds with tricks in each bed and sliders positioned in each trick.

Most machines use latch needles, and in operation the needle knits to the left, controlled by the slider in the left needle bed. The needle is transferred by moving it inwards so that the hooks at both ends of the needle are engaged by the two sliders. As the two sliders move outwards, one slider, pressed down by a cam, retains the needle. The right needle bed slider now has control of the needle and the yarn is fed in on to the opposite hook from the end used on the previous stitch. The slider on the right moves the needle to complete the purl stitch.

CIRCULAR MACHINES

These machines will produce knitted fabric of any reasonable diameter from 12 to 70 in. (30 to 177 cm). Circular machines are also used for knitting garment sections and for footwear manufacture. The latch needle is employed in the majority of machines.

The needles are placed in a circle and operate in a vertical direction within the cylinder. They are controlled by a disc which has small pieces cut out, leaving other parts to the original diameter. This system works similarly to the pegs on a dobby loom. The uncut area on the disc acts as a peg on the mechanism of the needles and causes the needle to work or not according to the type of machine. These discs are placed around the cylinder and the discs equal the amount of feeders in a given machine, each mechanical operation involving a feeder or Jacquard mechanism. A machine with thirty-six feeders in a two-colour design will revolve eighteen times and a three-colour design requires the machine to revolve twelve times.

In the machine operation, the needles move to a clearing height with the loops below the latches. A latch guard prevents the latches from closing, and while the needles move down, the old loop moves up the stem over the latch. The new yarn is fed on and as the needles move up, the latch is released and the new loop remains on the hook while the old loop is cast off.

With circular rib knitting the needles are placed in a vertical position in the cylinder, while the rib needles are placed horizontally in tricks in the dial.

WARP KNITTING MACHINES

In warp knitting machines the threads are looped length-

KNITTED FABRICS

ways through the fabric. Every needle in the machine has one thread, or two separate threads acting as one, to make a loop on one needle. This type of machine gives a high speed production and produces stable fabrics which are ladder resistant.

One type of machine employs the flying needle and in this method the needle rises and the tongue is withdrawn into the hollow stem of the needle, with the sinker in a forward position holding the fabric down. Yarn guides bring the yarn under the needles and over them so that the yarn is wrapped into the hooks. The tongue is raised to bridge the gap between the stem of the needle and the hook. The sinker moves across to assist with the loop and the needle moves down so that the new loop is drawn through the old loop which is cast off. During cast off, the sinker moves to hold the fabric down as the needle rises and while the tongue is withdrawing into the stem of the needle, the latter rises to begin the process again.

In most warp knitting machines, the lengthways threads are wound on to beams with one beam for each guide bar. The beam may be the full width of the machine, or may be in units of small beams, each a half or quarter of the length. The threads are fed in to the machine through a point bar and tension rail before being threaded through the yarn guides.

GAUGE

The term gauge is used to give an indication of the texture and quality of a knitted fabric. To produce a fine knit fabric, fine yarn and a fine gauge machine are used. A coarse knit fabric requires a coarser yarn and a heavier gauge machine. The gauge is not necessarily the number of needles per inch, but the relationship between the yarn and the particular needle size of the machine. A fine fabric can be 22 gauge, giving 1950 loops per square inch, and a coarse fabric would be in the region of 800 loops per square inch.

TYPES OF FABRIC

SINGLE JERSEY

A fabric used for cut and sewn garments in all types of yarn. It is produced by using a combination of plain knitting, tucking and floating, termed accordion, which gives fabrics with considerable stretch properties.

DOUBLE JERSEY

A rib structure fabric in which the stitches lie in two planes. The ribs produced in one plane close up to the other plane and form a double-faced fabric. The fabric stretches well and on the release of the tension will spring back into position. It is a very stable fabric with twice the elasticity of a plain knit fabric, but the stretch lengthways is negligible.

STRETCH FABRIC

A knitted fabric using a yarn which has been set as an untwisted or slightly twisted yarn in steam and afterwards twisted. Another method is by twisting a normal yarn, setting it by heat and untwisting it. Only yarns with good stretch properties are used and the stretch fabric is produced so that when the tension is released a loop distortion is formed. By stretching the fabric, the loop distortion is straightened out and the fabric expands. Another method of producing stretch fabrics is by knitting one course in an S twist yarn, followed by a knitted course in Z twist yarn.

TRICOT

A warp knitted fabric which is carried out by the transversal interlooping of the threads over more than one wale. Tricot is generally produced for underwear in fine yarns, including cotton, wool and synthetics. The fabrics tend to shrink somewhat in length and stretch in width.

KNITTED FABRICS

LOCKNIT

A warp knitted fabric with ladder resistant properties, made with two threads to every needle. The locking is made by the threads moving sideways to interlace. On the face of the fabric the wales have vertical loops, and on the back give the appearance of horizontal wales.

INTERLOCK

A double 1/1 rib fabric joined together with crossed sinker wales. The fabric is smooth on both sides of the cloth, and although it has a rib characteristic it does not unravel in the same way as a 1/1 rib. The loops on one face of the cloth are opposite to the loops on the other side.

ATLAS

A warp knitted fabric produced by using a double set of warp threads, which travel in opposite directions, forming laps for a number of courses. The direction of lapping is then reversed and the regular changes in direction produces a striped effect in the cloth.

MILANESE

A fabric similar in construction to Atlas, but produced by the interlooping over the whole width of the fabric by two threads moving in opposite directions. Milanese fabric shows a diagonal passage of two sets of warp threads from one side of the material to the other.

LACE

A variety of different fabrics are produced on Raschel warp knitting machines. These fabrics include all-over and banded laces, dress nets, curtain nets, veilings, etc. The machines employed are 12-bar, 14-bar, 18-bar and 24-bar, which allows a tremendous pattern scope.

Designing of lace is carried out in sketch form, preferably painted in white paint on black paper, and is to the size of the finished knitted lace. The sketch is then transferred to draughting paper by enlarging it by pantograph or magnification. There are several types of draughting paper lined in the form of hexagons. The size of paper used is the ratio of wales to courses and the relationship of the gauge of machine, allowing for shrinkage in the finished fabric due to dyeing or finishing. On draught paper the design is filled in, showing the movements of each guide bar at each course, and coloured crayons are used to plot these movements. The machine is threaded up according to the first course of pattern repeat as shown on the draught paper.

Modern knitted laces are formed using a basic net construction. For this type of construction, two guide bars are employed; both are fully threaded for all over laces and fully threaded in panels for banded laces. One set of warp threads is carried in one of the front two guide bars and forms knitted loops, and the other set of warp threads making the laying-in motions can be carried in any guide bar behind the knitting bar. The first warp threads are referred to as net-pillar and the second set of warp threads as the net-inlay. All the pattern threads are on beams at the rear of the machine.

Different types of machines are employed, depending on the complication of the design. In some lace production the machines have two needle beds and latch needles with the latches on the outside of the machine. The two sets of needles from both needle beds are in alignment with each other. Both needle bars are controlled from either one or two camshafts and the knitting action is performed vertically. The guide bars are swung collectively backwards and forwards between the needles and are controlled by continuously driven pattern chains. A thicker yarn is employed for the net inlay than is used for the net pillar. The lapping movements allow the laying in yarn to be connected to the ground structure; this depends on the draught used and may be an overlay or underlay within the structure. These

KNITTED FABRICS

laid-in yarns are not formed in knitted loops, but are laid in over or under half of the ground loops, and the traverse of inlay is dependent on the design motif.

ANGEL LACES

These are narrow lace trimmings produced on two-bar warp knitting machines. The narrow strips of fabric are connected together during the knitting process by cellulose acetate rayon threads. After the dyeing and finishing of the fabric the additional acetate rayon threads are dissolved in acetone to separate the lace bands. The yarns used to produce the lace are usually continuous filament synthetics and these remain unaffected by the acetone. Other forms of angel lace are produced on three-bar knitting machines and incorporate laid-in threads to form the design.

RIBBON SLOT LACE

With this type of lace, large holes, through which ribbon can be threaded, are produced by knitting together two lace strips. This combines inlay threads, net-pillar, net-inlay and straight pillar. Four needles within the width of the ribbon slot are left empty and the laying-in threads connect up the two panels of lace. Ribbon can be inserted into the laces as the knitting process takes place by special tubular guides which take the ribbon through the space created by the omission of a certain number of needles.

CURTAIN NETS

For curtain nets the machine used is similar to the lace machine, but a greater variety of widths are to be found than in lace manufacture. The machines incorporate needles with shorter hooks and the needle bar motion is reduced in width. In knitted net curtaining the yarns producing the pattern are less in number than those used for lace and are supplied from yarn packages at the rear of the machine. A uniform tension on the yarn is supplied by discs which allow for tension adjustment. Curtain nets are produced in a variety of yarns in the man-made fibres class, with a small percentage in cotton yarns.

MARQUISETTE

A fabric produced on two-bar or three-bar machines in which two or three full sets of warp threads are employed. The front warp threads knit continuously on the same needles, forming a straight pillar, while the other warp threads are inlaid into the pillar for a number of courses. After a certain number of courses, the inlaid threads make a large lapping motion and connect with the separate pillars together and form a small square mesh.

TULLE OR DRESS NET

Fabrics made with a simple net construction on machines which produce the material at a thousand courses per minute. The machine employs a latch needle with a very small hook which moves over a short distance. Two guide bars are used and the front bar knits a pillar construction while the back bar introduces the net-inlay threads. The two warps are accommodated on large diameter beams and the guide bars are controlled by pattern wheels. This type of fabric is used for veiling, dresses and millinery.

APPENDIX

Shuttleless Looms

There are a variety of machines working with different systems of weft insertion. The main feature they have in common is the insertion of the weft yarn into the shed from a large stationary package, which replaces the pirn in the conventional shuttle. This change in weft supply eliminates the pirn winding process. The various machines employ different methods to insert the weft, and the multiple gripper shuttle system has the widest application. Other machines use various rapier systems, water jet, air jet and the solid friction system.

In the two types of gripper shuttle looms, a much smaller shuttle than the conventional type is used, and both systems employ shuttle guides. With the multiple gripper machine, each shuttle is projected at a high speed through the shed and carries one warp thread. The shuttles return to their starting point by means of a conveyor belt. In the single gripper shuttle loom, the shuttle inserts one weft thread alternately from the left and right hand side of the loom. The shuttle guides are more widely spaced on this loom than for the multiple gripper system.

The rapier system looms fall into three categories, the single rigid rapier, the double rigid rapiers and the double flexible rapiers. In the first method the rapier, a long thin rod, enters the warp from the right, picks up the yarn from the weft package and returns through the shed trailing the weft thread. As the rapier has one wasted journey in travelling through the shed to pick up the weft yarn, the extra time taken up by this manoeuvre cuts down the loom speed.

With the double rigid rapiers the two rapiers enter the warp shed simultaneously, one from the left and one from the right. The left hand rapier carries the weft yarn to the centre of the warp and transfers the yarn to the right hand rapier, which pulls the thread through the shed to the selvedge. The double flexible rapiers system works on the same principle as the rigid rapiers loom, but the rigid rods are replaced by flexible steel or plastic tapes. These follow a curved path through the warp shed. The rapier system of weaving cloth allows a greater selection of colours to be woven in any sequence, and several weft colours can be inserted simultaneously, which is impossible on conventional looms.

Air and water jet looms work in the same manner. At each weft insertion a pump projects a jet of air or water from the nozzle. The yarn, lying in the nozzle, is accelerated through the warp by the jet. The water jet is a more efficient medium than air for accelerating the yarn, as the velocity of an air jet has to be ten times the velocity of the water jet. Air and water jets both disperse fairly quickly after leaving the nozzle and this limits the width of the looms. A solid friction loom involves two rollers rotating at high speed with the weft yarn between them. The friction between the rollers and the yarn accelerates the yarn across through the warp shed under its own inertia.

All the pirnless looms present a problem in weaving the selvedges. Each individual thread is a separate entity and precautions are taken to bind in the selvedge threads. There are several methods employed, and in one, a tuck-in selvedge is formed by each weft thread being tucked in to the fabric by the following thread. Another involves one end of each weft thread tucked in on the next but one weft thread. Other methods include a special binding thread woven into the selvedge on every pick, so that it is woven through the cloth. A hairpin selvedge is formed by pairs of weft threads forming a hairpin shape alternately facing from right and left. Leno is also employed on the last two warp threads to form a firm edge to the cloth. With other types, the weft threads are cut outside the selvedge and leave a fringe along the selvedges.

Pirnless weaving machines, particularly the water jet and rapier looms, are less noisy than the conventional looms. The loom speeds vary depending on the method of weft insertion. All the looms mentioned entail a high acceleration of yarn and this requires higher quality yarns to avoid breakages.

INDEX

Index

A

Accordion stitch, 206
Acetate rayon, 33
   dyeing, 183
   printing, 175
   washing, 196
Acid dyes, 187
Acrilan, 27
   dyeing, 184
   washing, 198
Acrylic fibres, 27
Air jet looms, 222
Alginate, 35
Alpaca, 18
Amazon, 87
Analysis of cloth, 64–67
Angel lace, 219
Angora, 16
Animal fibres, 11–20
Appliqued fabric, 69
Ardil, 35
   dyeing, 184
Asbestos, 36
Atlas, 87, 217
Azlin, 69
Azoic dyes, 189–190

B

Bag cloths, 69
Baize, 69
Bannockburn tweed, 83
Barathea, 79
Basic dyes, 188
Batik, 168–170
Batiste, 69
Bearded needle, 208
Beating, 56–57
Beaver cloth, 87
Bedford cord, 92–95
Beetling, 146
Billiard cloth, 69
Black Faced Highland, 13
Blankets, 79
Bleaching, 142
   cotton, 172
   jute, 181
   linen, 173
   silk, 174
   wool, 173
Block printing, 157
Blotch, 166
Bobbins, 55, 57, 110
Bonded fabrics, 138

Botany twill cloths, 79
   Quality yarn, 16
Boucle, 45
Box cloth, 79
Bri-nylon, 24
   washing, 197
Broadcloth, 70
Brocade, 111–113
Brocatelle, 113
Brushing, 145
Buckram, 70
Burning tests, 48
Butter muslin, 71

C

Cake and cop, 53
Calendering, 151
   Schreiner, 152
Calico, 70
Cambric, 70
Camel hair, 16
Camlet, 70
Canton, 70
Carding, 5
Care of fabrics, 192–202
Casement cloth, 70
Cashmere, 17
   washing, 195
Caustic soda tests, 49
Celafibre, 34
Cellulose Ester fibres, 33–35
Cellulosic fibres, 4, 30–33
Celon, 24
   washing, 197
Cheese cloth, 71
Cheeses, 52, 55
Chenille yarn, 45
Cheviot fleece, 13
   tweed, 83
Chevron, 85
Chiffon, 71
China silk, 19
Chintz, 171
Circular knitting machines, 214
Classification of breeds, 13
Cloud yarn, 44
Coir, 9
Combing, 5
Corduroy, 125
Cotton, 4–6
   dyeing, 180
   Egyptian, 4
   mercerised, 3, 143

Cotton *continued*
  physical appearance, 3
  printing, 172
  Sea Island, 4
  washing, 194
Count of yarn, 46–48
Coutil, 79
Courlene, 29
Courses, 204
Courtelle, 27
  washing, 198
Crabbing, 147
Crammed stripes, 73
Crash, 71
Crease resistance, 150
Creel, 55
Crepe de Chine, 71
Cretonne, 171
Crimplene, 26
  dyeing, 184
  washing, 197
Cropping, 144
Cross dyed cloth, 185
Cross weaving, 130–134
Cuprammonium rayon, 34
Curl or loop yarn, 45
Cuts, 46–48

D

Dacron, 26
Damask, 111
Denier of yarn, 46–48
Denim, 79
Degummed, 9
Desizing, 141
Dicel, 33
  washing, 196
Dimity, 87
Direct dyes, 187
  printing, 166
Discharge style, 167
Dispersed dyes, 191
Distorted weft, 95
Dobby loom, 90–92
  cloth, 94
  draft and peg plan, 93
Doeskin, 88
Donegal tweed, 83
Double cloth, 96
Douppion silk, 20
Drafting, 5
Drafts and peg plans, 57–62
Dralon, 28
Drawing, 5
Dress face finish, 144
Drills, 79
Dry cleaning, 199, 202

Drying of fabrics, 193
Duchesse satin, 88
Duck, 71
Dungaree, 80
Durafil, 31
  washing, 196
Duplex printing, 163
Dyeing of fabrics, 177–185
Dyeing of yarn, 50–53
Dyes, 186–191

E

Embossing, 152
Ends, 54, 59–61
  per inch, 54
Enkalon, 24
  washing, 197
Estamene, 80
Evlan, 31

F

Fair Isle, 205
Fancy or novelty yarns, 43–46
Fell of cloth, 57
Felt woven, 80
  non-woven, 139
Fibres, 2
  blended, 2
  extracted, 14
  filament, 2
  man-made, 21–35
  natural, 2–20
  staple, 2
Fibro, 32
Fibrolane, 35
  washing, 197
Filatures, 19
Finishing of fabrics, 141–154
Flake yarn, 45
Flame resistance, 151
Flannel, 80
Flannelette, 80
Flat bar machines, 211, 213
Flax, 6–8
Float stitch, 205, 206
Flock printing, 171
Fluflene, 26
Fluflon, 24
Fly needle, 208, 215
Folded yarn, 41
Folkweave, 97
Foulard, 81
Frieze, 80

G

Gabardine, 81
Galatea, 81
Gauge, 215
Gauze, 130–133
Georgette, 71
Gimp yarn, 44
Gingham, 72
Ginning, 5
Gilling, 15
Glass cloth, 72
  fibre, 36
  dyeing, 190
  washing, 198
Gloria, 72
Gossamer, 133
Grandelle yarn, 45
Gripper shuttles, 221
Grosgrain, 72
Guanaco, 17

H

Habit cloth, 88
Habutai, 81
Hackling, 7
Hanks, 50
Harris tweed, 83
Healds, 56–58
Hemp, 8
  dyeing, 181
Hessian, 72
Herringbone, 85
Holland, 72
Honeycomb, 97
  draft and peg plan, 98
Hopsack, 64
Huckaback, 98

I

Imperial cloth, 81
Interlock, 217
Ironing of fabrics, 194

J

Jaconet, 73
Jacquard, 101–110
  cards, 104
  fabrics, 110–118
  knitted, 205, 207
  loom, 105
  point paper, 101
Japanese satin, 88

Jersey fabrics, 206, 216
Jigger dyeing, 178
Jute, 8
  dyeing, 181

K

Kapok, 9
Kersey, 81
Knitted fabrics, 203, 216–220
Knitting
  machines, 208–215
  needles, 207–208
  stitches, 204–207
Knop yarn, 44

L

Laces, 217–219
Lampas, 113
Lap, 5
Lappet weaving, 134
Latch needles, 207, 208, 214
Laundry, 198
Lawn, 73
Lea, 46–48
Leno, 130–133
Limbric, 73
Lincoln and Leicester wool, 13
Line, 7
Linen, 6–8
  dyeing, 180
  mercerising, 144
  printing, 173
  washing, 195
Linsey, 81
Llama, 18
Locknit, 217
Lurex, 46
Luvisca, 88

M

Madras muslin, 132
  shirting, 88
Man-made yarns, 21–35
Marl yarns, 45
Marquisette, 132, 220
Matelasse, 115–118
Maud, 82
Melton cloth, 82
Mercerisation, 142–144
Merino, 14
Milanese, 217
Milling, 147

Mineral dyes, 190
  fibres, 36
Miralene, 27
Miralon, 24
Modacrylic fibres, 28
Mohair, 16
Moiré fabrics, 152
Moquette, 126–128
Mordant, 167
  dyes, 188
Mothproofing, 150
Mull, 73
Mungo, 14
Muslin, 73

Pile fabrics, 119–129
Pilling, 153
Pina cloth, 74
Piqué, 99
Pirn, 57
Pirnless looms, 221
Plain knitting, 204
Plain weave, 62–64, 68
Plastics, 136–139
Plated stitch, 208
Plush, 125
Ply of yarn, 41
Point paper, 59–63
Polyamide fibres, 22–25
Polyester fibres, 25–27
Polypropylene fibres, 30
Polythene fibres, 29
Polyurethane fibres, 29
Pongee, 74
Poplin, 75
Pre-metallised dyes, 189
Preparation of fabrics, 172–185
Printed and dyed styles, 166–171
Printing of textiles, 155–165
Printed warp, 170
Proteinic fibres, 4, 35
Purl knitting, 207, 213
P.V.C., 136
Pyjama cloth, 75

N

Nainsook, 74
Nap finish, 144
Needlecord, 125
Nep yarn, 45
Net curtaining, 130, 217
Net, dress, 220
Net-inlay, 217
Net-pillar, 217
Ninon, 74
Noils, 15
Nub yarn, 45
Nylon, 22
  dyeing, 184
  printing, 176
  washing, 197

O

Organdie, 74
Organzine silk, 20
Orlon, 27
  dyeing, 184
Ottoman, 74
Oxidation dyes, 180

P

Padder dyeing, 179
Paper fabric, 140
  yarns, 10
Peg plans, 62–67
Perlon, 24
Petersham, 74
Picks, 54, 61–64
  per inch, 55
Picking, 57
Piece dyed cloth, 185
Pigment dyes, 190

Q

Qualities of textiles, 1
  fleece, 11

R

Ramie, 9
  dyeing, 181
Raising, 144
Rapier system looms, 221
Raschel machines, 217
Raw silk, 19
Rayolanda, 32
Repeats, 110, 156
Reed, 56
Re-manufactured wool, 13
Removal of stains, 200
Repp, 75
Resist of reserve, 167–170
Retting, 7
Rib knitting, 205, 213, 214
Ribbon slot lace, 219
Rippling, 7
Romney Marsh wool, 13

Rotproofing, 151
Roving, 5
Rubberproofing, 150

S

Sanforizing, 151
Sarille, 32
   washing, 196
Sateen, 86
Satin, 85
Satinet, 88
S and Z twist, 42
Scouring, 141
Scrim, 75
Screen printing, 158–163
Scutching, 5
Seersucker, 76
Selvedges, 54, 204
Serge, 82
Sett of warp and weft, 54
Shafts, 56–64
Shantung, 76
Shearing, 144
Shedding, 57
Sheeting, 76
Shetland wool, 13
Shoddy, 14
Shropshire wool, 13
Shuttle, 57, 110
Shuttleless looms 221
Silesia, 88
Silk, 18–20
   dyeing, 182
   printing, 174
   washing, 195
Singeing, 146
Sinkers, 204, 209
Sisal, 9
   dyeing, 181
Size of Yarns, 46
Sizing of warps, 55, 141
Sliver, 5
Snaps, 48
Snarl yarn, 44
Solid friction looms, 222
Southdown wool, 13
Spanzelle, 29
Spinneret, 21
Spinning, 37–40
Spiral yarn, 44
Spot designs, 99
Spun bonds, 139
Spun silk, 20
Stains, removal of, 200
Staple, 2
Staves, 56

Steaming, 145
Stentering, 148
Stoving, 142
Straight bar machine, 209
Stretch fabrics, 216
Suffolk wool, 13
Sulphur dyes, 189
Swivel weaving, 134

T

Taffeta, 76
Tapestry, 113–115
Tartan, 82
Teklan, 28
   washing, 198
Terylene, 25
   dyeing, 184
   printing, 176
   washing, 197
Tex system, 48
THPC, 151
THPOH, 151
Ticking, 82
Tie and dye, 170
Tops, 15
Tow, 7
Towelling, 128
Tram silk, 20
Transfer printing, 164
Tricel, 33
   washing, 196
Tricelon, 34
Tricks, 207
Tricot, 216
Tuck stitch, 205, 206
Tuffle yarn, 44
Tulle, 220
Tussah silk, 20, 76
Tweeds, 82
Twill weaves, 78, 83–85

U

Ulstron, 30
Union cloths, 2

V

Vat dyes, 187
Velanizing, 150
Velvet, 119–123
Velveteen, 123–125
   finishing, 153
Velveteen, 123–125
   finishing, 153
Venetian, 89

Vicara, 35
Vicuna, 17
Vincel, 31
  washing, 196
Vinyon, 29
Viscose rayon, 30
  dyeing, 182
  printing, 175
  washing, 196
Vision nets, 133
Voile, 77

Weft knitting, 204
Winchbeck dyeing, 179
Wool, 11–13
  dyeing, 181
  printing, 173
  washing, 195
Worsted, 14–16
Woven fabrics, 54

W

Wales, 204
Warp, 54
Warp beam, 52
Warp knitting, 204, 214
Warping, 55
Warping order, 61
Water jet looms, 222
Waterproofing, 149
Weaving of fabrics, 56
Weft, 54

Y

Yarns, 2
  blended, 2
  count of, 46
  denier of, 48
  dyeing of, 50–53
  fancy or novelty, 43–46
  folded or plied, 41
  man-made, 21–36
  natural, 1–20
  S and Z twist, 42
  size of, 46
  tests on, 48
  twist of, 41